1840: Composition, or built-up, roofing introduced to the United States.

1845: Warren opens first composition roofing company in Cincinnati.

1847: Warren arranges to receive coal tar from Cincinnati Gas Light and Coke Co.—company gives its residue to him for free. (Thanks to Warren's experimentation during this period, tar, gravel, and felt rolls became dominant roofing materials.)

1848: Benjamin Barrett establishes roofing business in Chicago. An early employee is Moses Powell.

1850: Warren brothers begin refining coal tar.

1850: Lake Superior copper mines open in Michigan's Upper Peninsula. Helps copper's popularity.

1850: Roofing becomes specialized trade; breaks from carpentry industry.

1855: Benjamin Barrett hands business over to Samuel. He starts writing specifications and producing his own roofing materials.

1859: Moses Powell establishes roofing business in Chicago. He eventually will join Barrett.

1860: Introduction of prefrabricated buildings with corrugated, galvanized roofs for housing during Ca

1868: Michael Ehret pate system. One of first publis si-tion roofing specifications is produced.

1869: After Civil War, Powell and Barrett separate. Barrett's emphasis is on material production, Powell's on applications.

1871: Chicago Fire destroys 17,450 buildings, including manufacturing plant of Samuel Barrett, producer of "fireproof" composition roofing.

1883: Ebenezer Warren (brother of Samuel) and Michael Ehret form Warren-Ehret in Philadelphia.

1886: **Moses Powell and Samuel Barrett form Gravel Roofers Protective Association, forerunner of NRCA.**

1887: First annual Association meeting.

1890: Gravel Roofers reformulated and called National Association of Master Composition Roofers (also known as National Association of Master Gravel and Slag Roofers).

One Hundred Years of
ROOFING IN AMERICA

Researched and written by
John N. Vogel and Theodore J. Karamanski
with William A. Irvine
Loyola University Mid-American Research Center

Edited and compiled by
Christine Nolen Taylor
National Roofing Contractors Association

Published by the
National Roofing Contractors Association
Rosemont, Illinois

First edition
Library of Congress Catalog Card Number 86-062321
ISBN 0-934809-02-X
Printed in the United States of America

CREDITS

The National Roofing Contractors Association gratefully acknowledges the contributions of archival material from the following sources. The photographs and documents submitted by the following companies and individuals appear throughout the pages of this volume.

The Colonial Williamsburg Foundation
Virginia Division of Tourism
J. A. Piper Roofing Co.
Warren-Ehret-Linck Co.
Consumers/Dornin-Adams, Inc.
J. D. Candler Roofing Co., Inc.
The Norton Bros. Co.
The Eagle Group, Inc.
The Henry R. Fell Co.
Manville Corp.
N. W. Martin & Bros., Inc.
William R. Steinmetz, Sr.
Celotex Corp.
Schriber Roofing Co.
William C. Cullen
Jos. A. Sanders & Sons, Inc.
Nicholson & Galloway, Inc.
Kirberg Roofing, Inc.
F. J. A. Christiansen Roofing Co., Inc.
American Roofer magazine
The Lawson Roofing Co., Inc.
Mortenson Roofing Co., Inc.
Haws Roofing Co., Inc.
San Diego Convention and Visitors Bureau
CertainTeed Corp.
J. L. Adler Roofing Co., Inc.
Stanley W. Warshaw

Acknowledgements

The National Roofing Contractors Association would like to thank the following individuals for their contributions to this volume.

Library of Congress

State Historical Society of Wisconsin

Chicago Historical Society

Cincinnati Historical Society

John Crerar Library

Thomas Taylor and Mary Keeling, Colonial Williamsburg

Robert Vogel, National Museum of American History

Rosemary Estes, Museum of Early Southern Decorative Arts

John L. Ward, Ralph Marotta professor of free enterprise, Loyola University

Thomas J. Bennett, director of research services, Loyola University

Frank Cassell and Fred Olson, history department, University of Wisconsin, Milwaukee

Sheldon S. Cohen, manuscript reviewer

Richard Snyder and Sid Lewis, Asphalt Roofing Manufacturers Association

Edward Schreiber and Joan Boran, Construction Consultants, Inc.

Shirley Valade, J. D. Candler Roofing Co., Inc.

Thomas Cusa, Nicholson & Galloway, Inc.

Frank Lawson, Jr., The Lawson Roofing Co., Inc.

Michael Alcock, M. W. Powell Co.

Robert Linck, Robert E. Linck, Inc.

Centennial Committee, National Roofing Contractors Association

Beverly Kruger, chair, Centennial Committee

Fred Good, executive vice president, NRCA

William Good, former executive director, NRCA

Gale Kiesel, director, meetings and conventions, NRCA

Linda O'Brien, NRCA

Bea McSheffrey, NRCA

TABLE OF CONTENTS

Donald G. McNamara
President 1986–87

NRCA's Centennial provides us with the unique opportunity to study the evolution of our industry over the last 100 years. Through projects like this written history, we can learn about the inspiring tenacity and ingenuity of those who have preceded us. In doing so, we gain a healthier perspective on the current marketplace within which we must perform.

Anniversary celebrations are also morale builders and reminders to our customers of our many years of service. An anniversary can highlight our stability, dependability and competence. These qualities are inherent in our industry leaders of today—members of the National Roofing Contractors Association.

As professional roofing contractors, we dedicate this book to all those who will, over the years, experience the same sense of pride and wonder in reading about the development of this field that we experienced as it was compiled. We hope they will realize, as we did, that each event and discovery has helped lay the foundation for our continued growth.

We are, after all, only pinpoints in a vast universe of time. May we learn to honor the past, respect the present, and create the future.

Fred C. Good
Executive Vice President

Within our industry, as in all aspects of our lives, our daily actions form the next pages of history.

As major national and international events occur, we react to them. The record of these actions is our history. But history seldom records the planning, the dreaming, the "what ifs" and the "if onlys." In this way, history teaches us the importance of keeping our goals sharply focused and our sights fixed firmly on achieving results.

History also gives us valuable insight into the character and contributions of our industry's founders and leaders. The record of the roofing industry is the story of the men and women who had the courage, enterprise, individualism and cooperative spirit to accomplish their dreams.

This book chronicles a fascinating history. As long as the drive and ingenuity we have shown in the past continues, our future success is assured.

Ronald W. Reagan
40th President of the United States

I am pleased to greet members of the National Roofing Contractors Association on your 100th anniversary.

A century ago, this country began a period of progress that has given us the most productive economy and the highest standard of living in the world. You can be proud that your industry helped build homes, schools, and workplaces that are vital elements of this progress.

This standard of achievement continues as your Association's members are involved in preserving many of our Nation's most cherished landmarks. Your industry has applied the latest technology to the restoration of the Capitol in Washington and the Statue of Liberty in New York Harbor. I commend you on your dedication to professionalism, workmanship, and quality.

I offer best wishes for a successful celebration and a productive future. God bless you.

Despite shipping and quarrying problems, slate was successfully used on a number of important colonial structures, including the Virginia state capitol building in Richmond.

CHAPTER 1

The search for solutions: roofing in the New World

"There are some defeats more triumphant than victories." MICHEL DE MONTAIGNE

On May 13, 1607, three tiny ships lay at anchor on the James River. Packed aboard the vessels were a group of Englishmen and other Europeans intent upon establishing a settlement. Englishmen had tried before and failed—learning little of the American wilderness in the process. To sustain their settlement in the new land, the colonists brought a substantial amount of food stores. They feared the wild landscape would be unable to support them. Throughout the first day, supplies and colonists were ferried to land. By evening, 105 people had been put ashore on Jamestown Island.

That tiny settlement became the colony of Virginia, which marked the beginning of the development of the eastern United States. Other settlements and colonies followed, many of which shared Jamestown's trials of starvation, internal dissension, and fear of Indian attack. The colonists also shared the desire to implant in the New World the culture they had left in Europe. The habits, beliefs, and customs in Old England were familiar and comforting to pioneers isolated from the land of their birth. The colonists were also forced to face the reality of America: a harsh, rich environment filled with opportunity, but also demanding adjustment and innovation. The story of roofing in early America reveals the tension between the early settlers' instinct to employ Old World styles and the need to come to terms with the severe realities and abundance of the New World.

Roofing was not a specialized trade in early America, so the roofing system of the period must be studied from the perspective of the development of colonial building. That building tradition began not with the grand, red-brick colonial architecture seen today in historic Williamsburg and copied in posh suburban developments, but instead with rude dwellings that provided the most basic shelter. At both Jamestown and later in Plymouth Colony, the early colonists adopted, at least in part, the building form of the Indian. Conical-shaped wigwams built of branches, mud, wattle, and leaves were among the earliest homes. The roofs of such structures, though little more than an extension of the walls, were made of a weave of hazel brands stuffed with clay or daub.[1] Such covering did little better than keep the morning dew off its inhabitants. Captain John Smith, a chronicler of Jamestown's early trials, recalled that such roofs "could neither well defend wind nor rain."[2]

At Plymouth Colony, Pilgrims who tired of the leaky wigwams resorted to dugout shelters in the sides of hills. A wood frame was built to keep the earth from crashing in on the cavedwellers, but such a roof could not keep dirt from falling through the boards, fouling food and disturbing sleepers with thoughts of a cave-in. Even the pious Pilgrims, who saw themselves as "poor servants of Christ," could endure this type of rude shelter only temporarily.[3] For a society to develop in the New World, reliable housing stock was required.

It is natural that English colonists initially responded to the housing challenge by importing medieval building styles. Although great stone castles dominate the popular mental picture of the Middle Ages, the most common building element of that period was wood. Oak timbers were hewn by hand and erected into two-story frame cottages. But wood was dear in Europe. Access to forests for building material was carefully regulated by law. To minimize the expense of covering the frame with wood and the difficulty of working with heavy timbers, medieval English cottages were walled with either brick or, more commonly, a plaster of wattle and daub. English peasants covered their homes with thatch. The thatch-roofed and wattle-and-daub-walled homes of old England made their appearance in America only a few years after the establishment of a particular colony. To their owners, these homes represented security and comfort in a new land.

Yet a thatch roof provided only short-term security in America. In less than a year, the self-satisfied colonists in their tudor-style cottages found themselves facing either leaky roofs or, worse, the threat of fire. The wattle-and-daub walls lasted only slightly longer. The American environment, much harsher than temperate England, undermined the utility of medieval construction techniques in the New World. In Plymouth Colony, for example, the Pilgrims found a July-to-December temperature spread twice as great as that of Plymouth, England.[4] Hot summers and severely cold winters made wood-frame buildings expand and contract. Wattle-and-

Historians call the pre-1870 period "America's Wooden Age." Houses, fuel, even kitchen utensils came from the forest.

daub would yield to temperature in the same way as wood, so cracks and gaps often formed in the walls. Long winter freezes damaged thatch roofs, and summer heat and droughts turned them into fire hazards. In the growing towns of America such as Boston and New Amsterdam (New York), ordinances were passed in the 17th century to restrict the use of thatch.[5]

Faced with the need to find an alternative to thatch as a roofing material, the American colonists once more drew upon their medieval English heritage. During the late Middle Ages, many of the great manor houses used oak shingles. The high cost of wood kept this style from spreading widely among the poorer classes. However, in America, wood shingles became a very popular roofing option.[6]

What made shingles so critical to early American building was the abundance of wood in the eastern United States. From the Atlantic tidewater to the Illinois prairies stretched a vast and mature forest. Early explorers such as France's Jacques Cartier praised the new land as a "pleasant country full of all sorts of goodly trees, Ceders, Firres, Ashes, Boxe, and Willowes." Britain and Ireland had seriously depleted their forest resources by the 13th century. Although many original settlers were disappointed by the lack of gold and silver, they were as quick as Captain John Smith to see the potential of the forests. He wrote to England, "The treasures of this land have never been opened, nor her originals wasted, consumed, or abused." It was a land "overgrown with all sorts of excellent woods for the building of houses, boats, banks, or ships."[7] So important did wood become to American building, indeed to life in general, that historians of technology have dubbed the period before 1870 "America's Wooden Age." Not only was wood used to build most homes, but it was the source of material for construction tools, dinnerware, even cooking and heating fuel. From cradle to coffin, American life was based on wood.

Cypress, oak, and pine were all used to make wood shingles. But the most popular wood for shingles in early America was white cedar. Available generally in forests adjacent to low or swampy environments, white cedar enjoyed a reputation of being resistant to rot. During the 18th century, a specialized white cedar industry was established. So popular did this wood become that sizable cedar swamps in New England and New Jersey were completely stripped of trees. New Jersey cedar merchants tried to remain in business by marketing sunken trees salvaged from swamp floors. Many of these trees had fallen hundreds of years before Columbus, yet when they were floated to the surface, their wood was found to be in excellent condition. In fact, this ancient cedar was lighter and more durable than any other type of shingle. The wood fetched a very high price, and sunken cedar shingles were used on some of the most important roofs in early America, such as Independence Hall in Philadelphia.[8]

The making of wood shingles was vital to home construction. The method of construction was consistent throughout the United States and Europe. Until about 1840, even the same tools were employed.[9] The most important of these implements was the froe. The froe was a knife-like wedge, a tool with a blade affixed at a right angle to the handle. This was forced into a bolt of wood and driven by a club until the wood split from the back, forming a shingle.

The shingles then went through the additional process of being shaped on a shaving horse. There a draw-knife was used to fashion a type of wedge from the shingle, thinner at the top and thicker at the bottom. These shingles were usually 3 to 5 inches wide and 18 inches long. About 6 inches of the thick end, which was about a 1/2- to 3/4-inch thick, was exposed to the weather. The exposed end was frequently just squared off, but also frequently rounded. This way, the shingles were more resistant to curling and more aesthetically pleasing.[10]

The use of wood shingles is well documented, but perhaps no where as well as in Williamsburg. Evidence suggested that most of the wood shingles used there had come from the cypress tree, although yellow pine and white cedar shingles had been used, too. It is likely that oak and chestnut shingles may have been used also; those woods were not as popular as cypress or cedar.[11]

One early reference that dates the use of wood shingles to colonial Virginia is an excerpt from the *Act Directing the Building of the Capitol and the City of Williamsburg, 1699.* It states that the capitol " . . . shall be shingled withy [sic] cypress shingles . . . "[12] Although the legislators tried to use a different type of roof covering, they generally came back to wood shingles. A case in point was the governor's residence. The building specifications stated that the structure was to be covered with slate. By 1709, it had been decked and leaded; it was simply awaiting the arrival of the slate from abroad. When it did arrive, two-thirds of it had been broken and could not be salvaged. Caught in a bind, the Governor's Council directed that the roof be shingled instead.[13]

The popularity of the shingles extended beyond Williamsburg. Robert Beverly, a Virginian, observed that Virginia's " . . . common covering for Dwelling-Houses is shingle, which is an oblong square of cypress or pinewood . . . "[14] Shingle production in Virginia also took on an international flavor in the first half of the 18th century when the colony began exporting the material to Madeira, the West Indies and Barbados.[15] Notable buildings in the North that used wood shingles included the first New York state capitol in 1809.[16]

Popular as shingles were, their availability was limited to the number that could be produced by the human hand. This limitation was removed very early in the 19th century with the invention of the shingle machine. An advertisement in an 1802 edition of the *Maryland Herald and Eastern Shore Intelligencer* described the machine: "NEW INVENTION. Shingle Machine invented by D. French of Connecticut. 2 stroke process—1st one shaves shingle, 2nd one joints it."[17] The innovations kept coming and by 1809, shingle machines were being developed that could be powered by wind, water, steam engine, horse or manual labor.[18] By 1850, at least 20 patents had been issued for new shingle machines or various aspects of the shingle manufacturing process.[19]

Another important early wood roof was made of clapboard. As used in Colonial Williamsburg, this type of roof consisted of narrow oak boards about 4 feet long.[20] Clapboard is a uniquely American contribution to the building industry.[21]

Initially, clapboards were split off logs by hand. Using a mallet and a froe, logs were split repeatedly until the boards remaining resembled a wedge, very thin on one edge and 3/8- to 1/2-inch thick on the other. This method of clapboard production was followed by one using a pit saw. To employ this technique, one man stood above a log and another in a pit below. Together, with a whipsaw, they would saw a log into clapboards using an up-and-down motion. The whipsaw in this method was soon replaced by a "framed pit saw." As its name implies, this saw was held within a frame that the men then moved up and down, allowing a thinner, more efficient sawblade to be used.[22]

Although clapboard and shingle were the most common type of wood roof, other types were also probably used. Evidence suggests, for instance, that a small Episcopal church in northeast North Carolina may have originally had a board-and-batten roof. Furthermore, this particular roof seems to have been coated with tar to preserve the wood.[23]

The use of pine tar on wooden roofs was not uncommon. It is one of nature's laws that wood (thus wood shingles, clapboards and board-and-batten) rots when exposed to the weather. To combat this deterioration, preservatives were applied to the wood roofs. Very often the preservative was pine tar.

The Governor's Palace in Williamsburg, Virginia was supposed to be roofed with slate, but two-thirds of the shipment from England was damaged. The Colonial Council decided on wood shingles instead.

Pine tar was an abundant commodity. British legislation in 1704, 1722 and 1748 offered a financial inducement for its use and aided its production in the colonies. The British had a keen interest in pine tar, as it was used extensively to maintain the watertightness of its navy's ships.[24]

Harvesting pine tar was an important industry in early America. Woodsmen would build a kiln in the forest. Dead wood was collected and cut into 2- or 3-foot lengths and slowly burned. The key to the process was a hole in the middle of the kiln, which was linked by a conduit to a receptacle (usually a prepared ditch, sometimes a large barrel). The wood was allowed to smolder for days, burning slowly downward through the pile. The slow, steady heat gradually drew the tar out of the wood; it dripped into the conduit and then into the bucket. Depending on how large the pile of pine was, this method could produce a significant amount of tar, often as much as 10 barrels a day. The simplicity of the process made it available to any farmer who had access to pine trees.[25]

It is uncertain when pine tar was first used as a preservative for wood roofs, but it is likely that, based on historical references about the Poles making tar and pitch, it dates back to the early Jamestown settlement. Dated references to the use of tar on roofs have been found as far back as 1679.[26] Bryton Episcopal Church in Williamsburg, completed in 1683, required two 28-gallon barrels of tar to cover the shingles.[27] Once tarred, roofs had to be re-tarred to maintain the level of preservation for the wood. In Williamsburg, the Governor's Council resolved in 1705 that "the roof of the capitol and prison be tarr'd again this summer as often as the overseer of the building shal think fitt." Additionally, a 1765 building lease specifies that the "shingles [are] to be tarred once in every two or three years."[28]

Tar was not the only material used to preserve wood roofs from the weather. They were also painted, sometimes with fish oil mixed in. A substance known as Egyptian Cement was also used to protect some wood roofs. The June 28, 1811 edition of the *Raleigh* (North Carolina) *Star* included an advertisement with a recipe for Egyptian Cement that "will resist water, and some persons imagine fire . . . " It called for a common pail to be filled with lime and mortar, to which a pound of brown sugar was added.[29] It is presumed that water was then added to the combination and mixed.

The matter of fire focuses on one of the drawbacks of the various wood roofs discussed. Though they were not as readily flammable as thatched roofs, wood roofs were certainly combustible. Wood roofs were outlawed in some major cities, but that was not their death knell as it was for thatched roofs. Instead, many concoctions were devised to make wood roofs fireproof. At their simplest, these mixtures consisted of paint mixed with oil, lead, tar and ochre or black lead.[30] The more elaborate versions included the following recipe for fireproofing a wood shingle roof from "an old citizen" of Boston in 1819: "Mix one ounce sulphur, one ounce red ochre and six ounces copperas together, and make a powder of them. After having prepared the roof . . . with some glutinous liquid, spread or strew this powder thereon; and repeat it three or four times. It will soon become hard; and will fortify the wood effectually against fire."[31]

There was also a recipe for French Cement: "Take as much lime as is usual in making a full pail of whitewash and let it be mixed in the pail nearly full of water; in this put two pounds and a half of brown sugar, and three pounds of fine shale, mix well together, and the cement is completed."[32] Paint additives as fire retardants date back to the early 18th century, at least; these latest compositions were all early 19th century. Another late 18th century entry came from the Royal Society of Sweden. It was picked up in the United States by the *North Carolina Minerva and Fayetteville Advertiser* on December 3, 1796. It suggested that: "After roofing a house with wood, boil tar in a kettle and mix fine pulverized charcoal with it, till it is the thickness of mortar; spread this with a trowel about a fourth of an inch thick over the roof; it will soon grow hard and defy all vicissitudes of weather. Roofs thus covered have stood in Sweden above a century, and still want no repair."[33]

A way to circumvent the problem of fireproofing a wood roof was to use a completely noncombustible material, such as slate. Archaeological excavations at Jamestown suggest that slate roofs may have been used there as early as 1625.[34] The use of slate in the colonies can definitely be dated to 1654 Boston. It was quarried locally from Slate Island and Hangman's Island in Boston Bay.[35]

The use of slate grew slowly at first. A house in Philadelphia, built by Samuel Carpenter between 1687 and 1699, was covered with slate. Within the vicinity, it was known simply as the "slate roof house," implying that slate was still infrequently used.[36]

One of the major problems that retarded the early growth of slate was illustrated in the story of the slate shipment for the governor's Williamsburg home: it was a very difficult and expensive commodity to transport. At the outset, American slate did not weather the extremes of the New England climate well. Therefore, much of the early slate used was imported from quarries in Wales.

Its transportation was an expensive and tenuous proposition, as demonstrated by the damage travel inflicted upon the slate destined for the Virginia governor's residence. Robert Beverly observed that " . . . tho' they have slate enough in some particular parts of the country . . . neither has anyone thought it worth his while, to dig up the slate, which will hardly be made use of, till the carriage there becomes cheaper and more common."[37]

One of the reasons slate was so difficult and expensive to transport was its weight. Because it was so heavy, sturdier roof frames were needed on the buildings that were going to support it. This created yet another cost that added to the overall expense of using slate.

In spite of these initial problems, slate did become a popular roofing material. It was used on such highly visible buildings as the Old State House in Boston and the Virginia Capitol Building. Before the 18th century ended, both New York City and Boston had ordinances that recommended its use, along with tile, as a fireproof roof covering. In fact, more than half the roofs in New York City were thought to be covered by slate by 1830.[38]

As the demand for slate grew, more and more American quarries producing quality slate opened. There were three major slate-producing regions in early America. The first one was the New York/New England area. Starting with the commercial development of the Hangman's Island quarries in 1721, quarries were opened in Lancaster, Massachusetts in 1750, New York in 1810, and Vermont in 1812. Another region was the Pennsylvania/ Maryland border, where quarries were opened in 1734 and 1808. The final region was in the Virginia/Maryland area. Three quarries opened up in Buckingham County, Virginia along the James River between 1797 and 1804. Three more opened in the Baltimore area between 1809 and 1815. In 1812, one opened up along the Susquehannah River in Virginia and in 1817, a new one in Buckingham County.[39]

Quarrying slate was a very labor-intensive operation. As a quarry was opened, the top layers of old, weathered stone had to be removed, as they were not of the quality needed. Once the good stone was reached, huge blocks were broken out, layer by layer, and lifted to the side of the quarry by a hoist, where they were dressed by hand. The sculptor/blockman split the blocks into sections 2 inches thick. An artisan then split the 2-inch blocks in half repeatedly until they were the desired thickness, usually about 3/16 of an inch. The trimmer cut the slates to the prescribed size.[40]

By the mid 1800s, there was a very large demand for domestically produced slate. Several factors combined to create this situation, not the least of which was the increase in the American population. It tripled between 1830 and 1860, and there was a corresponding increase in the demand for slate. Because the construction of railroads and canals reduced the cost of transportation, the price of slate declined. Finally, as architecture was developing into an art, the roof was seen as an integral part of a building's overall design. Because slate was an extremely attractive roof covering, it merged well with many of the new architectural designs.[41]

It is said that the sincerest form of flattery is imitation. As early as 1774, attempts were made to copy slate. Speaking of the houses in Fredericksburg, Virginia, a visitor in that year noted that ". . . all [were] covered with wood shingles made in the form of slates about four inches broad, which when painted blew you wou'd not know it from a house sclated with Isedell Sclate."[42] In 1806, a gentleman named Thomas Woodsend even patented a method of producing artificial slate. Though it is not known what success he had, he did attempt to capitalize a company to produce the material commercially.[43]

Tile was another fireproof roof covering. It appealed to the consumer of the day because it was a very durable material that required little maintenance, and it was a very poor temperature conductor. As with slate, tile's popularity received an early push when it was recommended for use in the building ordinances of Boston and New York.[44]

Early tile was imported from Holland, but it was not long before the need to manufacture tile domestically became apparent. Early tile production in America dates back to the 1650s, with several yards in operation by the time of the Revolution. As the demand for tile grew, so did the tile yards. In 1800, a tile factory opened in Charleston that was to supply the Atlantic Coast region.[45]

Tile's popularity did not carry very far into the 19th century. Nothing could compete with the low cost and availability of wood shingles. Because wood shingles had such appeal, the only natural constituency tile had was within designated fire districts. After 1810, there was even competition from slate and other new fireproof materials. Another almost inexplicable problem with tile roofs was the criticism, voiced several times during the 1830s, that tile roofs looked clumsy and unattractive. Unlike slate, which was achieving great acceptance by architects, it seems few people wanted a house with a "clumsy-looking" roof.[46]

The new fireproof roof coverings that helped drive tile out of vogue were made of metal. Perhaps the earliest metal used to cover a roof was copper. Sheet copper had long been used as a sheathing for the hulls of sailing ships; its water-resistant qualities were well known. Copper's use on American buildings can be traced at least to 1763-64 when material imported from Bristol, England was used on the New York City Hall. Copper was subsequently used on the Maryland State House in 1774, the First Bank of the United States in Philadelphia in 1796 and Bulfinch's State House in Boston in 1802. Despite its use on these highly visible buildings, copper did not achieve popularity until the Lake Superior mines of Michigan's Upper Peninsula were opened in the mid-19th century.[47]

Zinc was still another of the new metals used on roofs. Although zinc plates were first rolled in Sheffield, England in 1805, the industry later developed in Belgium. Zinc was a new material and there was no ready market to embrace it. A market had to be created, so in 1811-12, several buildings in Belgium were roofed with zinc to demonstrate what a wonderful roof covering it would be.[48]

It did not take long for the enthusiasm zinc created to reach America. On January 4, 1816, an advertisement appeared in the *Federal Gazette and Baltimore Daily Advertiser* that told of how zinc "can be made use of in cases where lead, tin and copper are employed; such as covering terraces, and houses . . . "[49] Two months later, John Bouis, a Baltimore tinsmith, was advertising that he had zinc spouts and gutters on display, and that zinc " . . . is warranted to last ten times as long as tin, but with very little more cost."[50] Further up the coast, New York had also discovered zinc. It was reported that in 1816, several houses had been covered with zinc, including "the college [Columbia] of that city." In July of 1818, a Richmond, Virginia newspaper contained an advertisement that

Tile was first manufactured in the United States in the 1650s, but its popularity declined in the 19th century as other fireproof materials, most notably metal, became more cost-effective.

heaped lavish praise on zinc. "It has been sufficiently ascertained by experiment, both in this country and Europe, that this valuable invention has introduced a better covering for houses than any other materials that have heretofore been in use." The advertisement also said zinc was "long lasting, solders well, does not need painting, assures pure water [for the cistern] and is one-half the cost of copper."[51]

Zinc was still the subject of some controversy. Its popularity in America was cyclical. Introduced around 1816, it was out of fashion by the 1840s, only to be revived later in the century.[52]

Few roof covering materials have generated the heated debate that zinc has. One such exchange occurred between a Dr. Gale of New York in *The Mechanic's Magazine* and Professor A. Caswell in the *American Journal of Science and Arts* in 1836-37. Gale submitted that zinc was difficult to make watertight, tainted the water diverted to a cistern, and offered little resistance to fire because of its low (700 degrees Fahrenheit) melting point.[53] Caswell retorted by saying that it was not singularly difficult to make watertight; he dismissed the argument over tainted water because experiments he conducted showed no contamination; and he said a 700-degree melting point offered just as much protection as any other metal. But if the concern of fire was still not put to rest, Caswell suggested covering the deck with a heavy coat of lime mortar, then putting the zinc over it.[54] Each of these arguments was lacking in substance, but they do demonstrate the emotion the material generated.

The final and perhaps most universally accepted and timeless roof covering to come out of this period was tinplate. One of the early American buildings covered with it was the Exchange Coffee House of Boston, built in 1808.[55]

Man has known of tin since 3500 BC, first as a co-ingredient with copper to make bronze, and about 1,000 years later, as an independent substance. Tinplate, the actual predecessor to roofing tin, was first made in Bavaria in the 14th century. England gave birth to and nurtured a fledgling tinplate industry with protective tariffs in 1703 and 1706 that levied duties on all tin imported to the island.[56]

Wrought iron was the first material to be tinned, thereby creating tinplate, and it had to be hammered into sheets by hand. This system was soon improved with the introduction of a water-driven hammer called a helve. Adapting a machine designed in 16th-century Italy to flatten soft metals for coin stamping, rolling mills were used to form the wrought-iron plates for tinning by 1730. The use of the rolling mill to form the plate for tinning was hailed as " . . . a technological development of the first order."[57]

Reduced to its simplest level, tinplate is produced by dipping a sheet of iron into molten tin and removing it. In practice, however, tinplate production was very difficult, because the sheets of iron had to be immaculate and the bond between iron and tin had to be very tight to assure a quality product. After the introduction of rolling mills, the process was further perfected when the grease pot was introduced in 1745. The grease acted as a flux and made the tin and iron adhere more strongly. In 1760, cleaning the sheets was made much more thorough with the use of mineral acids in the "descaling and pickling" stages.[58]

Meanwhile, the industry in England began to grow. The tariffs originally had little effect on tin production. By 1750, there were only four tinplate works in the nation. A dramatic growth period followed, and that number jumped to 32 in the ensuing 50 years.[59]

The actual method of British tinplate production in 1818 was fairly complex. The wrought-iron sheets were rolled in the rolling mills. They were then cleaned, or scaled by being placed in a heated solution of muriatic acid and water. Then they were cold-rolled, resubmerged in an acidic bath for 12 hours, scrubbed with hemp and sand and stored in clean water. When the time came to tin them, they were submerged in the grease pot (the flux to facilitate adherence of the tin) and then in a pot of molten tin for 90 minutes. When taken from the tin pot, the sheets were washed and brushed with hemp again and given a final dipping in the tin pot.[60]

The development of the tinplate industry in Great Britain, and the manufacturing procedure that followed, is of particular interest to Americans, because when tinplate was introduced as a roof covering here, all of it was imported from England. In fact, the American tinplate industry did not begin until 1890.

The story of tinplate roofing in the United States began in the colonial period. Like other materials used in those early years, such as thatch, wood shingles, tile and slate, tin was destined to be one of the materials that launched the 19th century roofing industry.

J. D. and George Candler proudy open their Detroit ironwork shop in 1884. The firm would soon expand to include slate, tin and iron roofing work.

CHAPTER 2

The decision to specialize: the origins of the roofing contractor

"Make it thy business to know thyself, which is the most difficult lesson in the world."

CERVANTES

Today the application of roofing systems is an important specialization in the contract construction industry. Yet the roofer did not take his place as a unique building specialist until the middle of the 19th century. Carpenters dominated construction in early America. The emergence of the specialized contractor from the generalized builder is an important chapter in the history of roofing.

Men skilled in building were among the most important settlers in colonial America. An early Bostonian summed up the situation succinctly when he noted that a man who "builds a good house to defend us from windward weather, is more serviceable than the curious carver, who employs his Art to please his fancy . . . what's more substantially serviceable to Mankind, is much preferable to what is less necessary."[1]

The experience of the earliest English colonies underscores the importance of skilled construction workers. At Jamestown, the first colonists were adventurers seeking fortune instead of men capable of building a settlement. Captain John Smith remarked that the nobles present "never did know what a day's work was," while the other settlers were "poor gentlemen, tradesmen, serving men, libertines, and suchlike, ten times more fit to spoil a commonwealth, than either bring one to life, or help maintain one."[2] Ill-housed and ill-fed, many colonists died trying to build the Virginia Colony. The decision to import skilled workers to America for constructing permanent dwellings was critical to the colony's eventual success.

The hazardous nature of travel to and life in the New World made English builders leery of relocating there. Builders were often in short supply where they were needed most, in new and growing settlements. Jamestown, for example, was forced to import Polish artisans to build that flagging colony. The first five Poles arrived at Jamestown in October 1608. Upon surveying the pathetic settlement, Zbingew Stefanski commented: "That first year must have been unbelievably difficult for the Englishmen. They did not deliver enough lumber for building houses; they did not produce any pitch, flack, or glass; however, quarreling and fighting among themselves was prevalent . . . "[3]

Skilled in construction arts, the Poles made an immediate improvement in the colony. To turn the abundant forest into a resource, they built a saw mill and began to cut "beams and planks without respite." Among their most important services was the installation of wood shingle roofs, the first in North America, on the frame dwellings they built. The shingles proved superior to the bark or haphazard thatch roofs Jamestown's amateur adventurers had put up earlier. In this respect, the Poles of Jamestown were the forerunners of the roofing contractors of today.[4]

The Poles in Jamestown were the first skilled craftsmen and builders to arrive in the country. Slowly, as more and more skilled craftsmen perceived that the opportunity of the New World outweighed the hazards, and as others were invited for specific jobs, they came across the sea to settle. These were men knowledgeable in the art of creating substantial and durable buildings.

For the colonial craftsmen, finding a steady and reliable market within which to ply their trade was essential. Most often such a market existed in a village or town. Unfortunately, those arriving in the South found a more difficult situation. The South was becoming a very rural, agrarian society. Tobacco, rice and indigo were the main crops grown on large plantations, some of which rivaled the English country estates and manors in both size and grandeur. Because of the overwhelming size of the plantations and the restrictions they inflicted on land use, as well as the fact that they were largely self-supporting, small towns generally were not important in the early South. As a result, the skilled builders were deprived of their natural market. Housewrights led a very nomadic life, moving from one estate to another for jobs.

The situation worsened for these craftsmen with the passage of time. Plantation owners frequently assigned their slaves to the builders for use as laborers. The slaves began to learn the trade, and, because slave labor was cheaper than skilled labor, the housewrights and builders were often replaced by slaves. In fact, some plantation owners began hiring out their slaves, thereby giving the craftsmen local competition. Because of competition from slaves, and uncertainty about where their services could be sold, many craftsmen were obliged to farm for a living.[5]

The Southern craftsman's condition began to stabilize as more and more smaller merchants and farmers gained footholds in the region, and provided the stable market that the craftsman needed to survive. In fact, some towns composed mainly of craftsmen began to develop in the South—towns like the Moravian settlement of Salem, North Carolina.[6]

The Middle Colonies and New England also developed an agrarian society, but unlike the sprawling plantations of the South, their farms usually were smaller and more oriented to nearby towns. These towns and the neighboring farms provided the necessary market that the craftsmen in the South had so sorely missed. Each town usually had a blacksmith, carpenter and housewright. And because there were about 550 towns in New England by 1776, a substantial market existed for these crafts.[7]

Life in a small town was not demanding enough to require specialization within general crafts. It was not until towns grew and became more urban that the growing marketplace and an increasing labor supply dictated necessary specialization.[8] It was a way of trying to insure one's economic survival by carving out a specific new market. For instance, a town might have had 20 carpenters, all of whom were competent in cabinetmaking. But if one carpenter decided to call himself a cabinetmaker, he would very likely attract much of the town's cabinet work. He would have stood out from the other 19 carpenters and probably would have had little competition until another one decided to try the same thing. Also, the use of new building materials required specialization within the building craft. Masons, bricklayers, plasterers, painters and plumbers were among the craftspeople that evolved from this type of innovation.

Despite the wave of specialization within the construction industry in the 18th century, there is no suggestion that a roofing trade was emerging. Instead, evidence suggests that carpenters and housewrights applied the various types of wood roofs, and tinsmiths, coppersmiths and other metal workers applied metallic roofs.

Carpenters had been charged with applying wooden roofs for many years. When the public hospital was built in 1771, the contract between Benjamin Powell, carpenter, and the City of Williamsburg, Virginia specified that " . . . the whole roof [was] to be covered with plank and good cypress shingles . . . "[9] In 1856, a contract between Chicago developers and their respective carpenters, James Watson and J. Campbell, called for the carpenters to apply the roofing on their new buildings on Lake Street in Chicago.[10] In fact, as late as the 1870s, carpenters' duties often included the installation of roofs.[11]

The addition of roofing to the jobs of tinsmiths and coppersmiths is an interesting one because, unlike carpenters, they were not really included in the building industry. Instead of constructing buildings, they were producing stills, and a variety of made-to-order kettles and cooking materials: brew kettles, fish kettles, tea kettles, wash kettles, hatters' kettles, fullers' kettles, coffee pots, chocolate kettles, dripping pans, sauce pans, tin plates and dishes, bake pans, tin ovens, sheet-iron stoves, cabin stoves, stove pipes, and scales and weights.[12] But because the materials of their trade were used for gutter, downspout and roofing work, they were the only ones familiar enough with the material to successfully address the new uses in spite of their lack of construction experience.

One such firm that expanded to include gutter and roof work was founded by Christopher Raborg of Baltimore. The Raborg Company makes an interesting study not only because the firm ultimately dealt with metal roofs, but because Raborg was a civic-minded craftsman with a distinct sense of professionalism. Born in 1745, Raborg was in business for himself as a coppersmith and tinsmith by 1785. At that time, his product line included stills, kettles and other standard copperware and tinware. By 1802, his son, Christopher, Jr., was included in the business and the firm's name appropriately changed to Christopher Raborg and Son. The newly formed partnership added a brass line in 1804 and continued to offer the traditional line of copperware until the death of Christopher, Sr. in 1815.

Judging from the numerous advertisements for apprentices and journeyman tinsmiths and coppersmiths, Raborg ran a very successful business. Yet he was also able to maintain a very active commitment to the community. In 1789, he was named manager of the High German Reformed Church lottery. His professionalism within his trade was recognized when he was elected manager of the Baltimore Mechanical Society in 1793. His fellow citizens elected him councilman from the Fourth Ward in 1802 and in 1803, he participated on a Baltimore grand jury.

It took five years to clear Raborg's estate and settle all claims and accounts of Christopher Raborg and Son. When that was accomplished, Christopher, Jr. continued the business as " . . . Surviving Partner and Successor of the late firm of Christopher Raborg and Son."[13] One of the first advertisements for business he placed as successor to his father's firm appeared in 1820. This ad is significant, because it specifically mentions the lead on hand that was "suitable for the roofs of buildings."[14] Clearly, the Raborg concern had by then specifically acknowledged the use of its material for roofing. No mention was made of actually applying it; Raborg seems to have been doing more of the work of the supplier than the actual roofer. In 1822, Raborg ran an advertisement that read, in part, "Copper and Tin Spouting for Houses, put up in the most substantial and durable manner . . . "[15] Thus work on roofing appurtenances was acknowledged and the Raborg firm found itself in a facet of the construction business as well as the domestic tinware and copperware business. Raborg was among the earliest roofing contracting firms in the United States.

Hugh Bonner is another example of an earlier roofing businessman who applied metallic roof coverings. An advertisement in the November 14, 1806 edition of Baltimore's *American and Commercial Advertiser* noted that Bonner was spending most of his time "laying copper on buildings and forming gutters . . . "[16]

An important factor in the specialization of construction tasks into distinct trades was the mid-19th-century revolution in building techniques. To keep pace with the burgeoning population of the United States, standardized and streamlined building components became basic to the construction industry. The balloon frame was among the most important of the new inventions. Chicago builder Augustine Taylor used a wood skeletal design to build an inexpensive wood church in 1839. Lightweight dimension lumber, held together by machine-produced nails, made this style of construction easier, cheaper, and quicker than the old mortise-and-tenon system.[17] This technique allowed less skillful men to put up buildings in much less time. Thus, more buildings were built and more roofs were needed. The old housewright who had the skills to build a house from start to finish was replaced by the specialized contractor.[18]

Slowly, as specialization within the building industry became more prevalent, some people began calling themselves roofers. Charles S. Parker of Boston, for instance, identified his profession as housewright from 1842 to 1845. From 1846 to 1848, he was in business with a Mr. Palmer. They were listed as "Carpenters, Coppers and Zincers" in the 1846 *Boston City Directory*, and as metal roofers in the 1847 directories.[19] In 1848, Parker went into business with Gillian B. Wheeler and they worked together as housewrights and coppers until 1851. The partnership was dissolved in that year and Parker continued to work independently, identified specifically as a metal roofer. Wheeler continued work, calling himself a carpenter through 1854, when he added roofing to his list of skills.[20] Other Bostonians calling themselves roofers included Charles Fay, a tin worker

An 1868 advertisement for the Ehret slag roofing system lists references and informs readers that "state and county rights" are for sale.

from 1845 to 1848 and a tin roofer after 1849. John Sargent was identified as a carpenter from 1848 to 1851 and thereafter as a metal roofer.[21]

An interesting result of the growing specialization in roofing was a crossover of skills. Carpenters sometimes laid metal roofs; conversely, city directories indicate that metal roofers would occasionally put down roofs of wood shingles. The reason for this, undoubtedly, was that when offered work and the chance to make money, the newly specialized roofer didn't want to turn it down. As a result, some contractors learned how to work with the new material to the extent that it became part of their product line. During economic difficulties, the roofer with the widest variety of products, i.e., composition roofs, metal roofs, slate roofs and shingle roofs, had the greatest opportunity to obtain work.

In Philadelphia, Richard R. Cumming was the first contractor to call himself a roofer. Listed in the city directories as a tinplate worker from 1835 to 1836, he shifted his occupation to hardware from 1837 to 1841. In 1842, he identified himself as a metal roofer and continued as such for many years. George Johnson was a tinsmith from 1835 to 1855, but during that period so much of his work was in roofing that he began to identify himself as a metal roofer.[22]

Many of the people specializing in early roof-

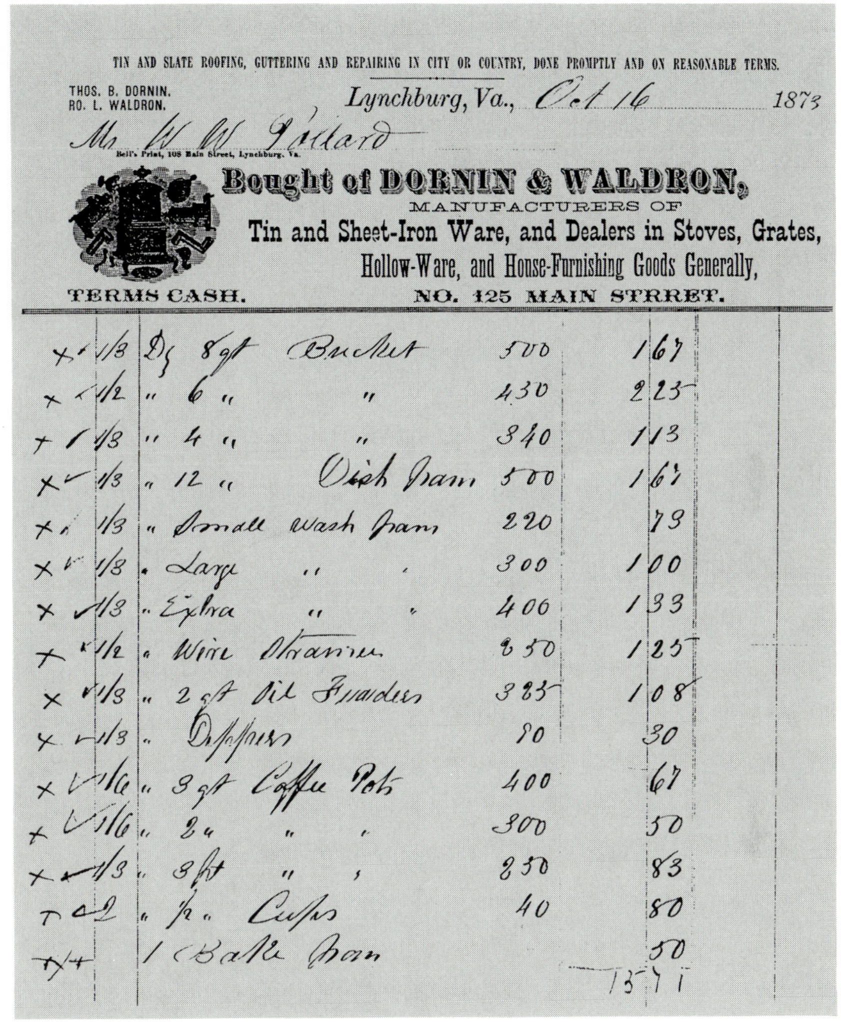

TIN AND SLATE ROOFING, GUTTERING AND REPAIRING IN CITY OR COUNTRY, DONE PROMPTLY AND ON REASONABLE TERMS.

THOS. B. DORNIN.
RO. L. WALDRON.

Lynchburg, Va., Oct 16 1873

Mr W W Pollard

Bell's Print, 108 Main Street, Lynchburg, Va.

Bought of DORNIN & WALDRON,
MANUFACTURERS OF
Tin and Sheet-Iron Ware, and Dealers in Stoves, Grates,
Hollow-Ware, and House-Furnishing Goods Generally,

TERMS CASH. NO. 125 MAIN STREET.

1/3	Dz 8 qt Bucket	500	1	67
1/2	" 6 " "	430	2	15
1/3	" 4 " "	340	1	13
1/3	" 12 " Dish Pans	500	1	67
1/3	" Small Wash Pans	220		73
1/3	" Large " "	300	1	00
1/3	" Extra " "	400	1	33
1/2	" Wire Strainers	250	1	25
1/3	" 2 qt Oil Feeders	325	1	08
1/3	" Dippers	90		30
1/6	" 3 qt Coffee Pots	400		67
1/6	" 2 " "	300		50
1/3	" 3 pt " "	250		83
2	" ½ Cups	40		80
	1 Bake Pan			50
			15	71

An 1873 invoice from Dornin & Waldron, dealers in "tin and sheet-iron ware." Founded in 1861 and now Consumers/Dornin-Adams, Inc., the firm is one of the oldest roofing and sheet metal contracting operations in the South. The bill notes such expenses as a dozen 3-quart coffee pots for 67 cents.

J. A. Piper began his career in roofing and sheet metal in turn-of-the century Durham, South Carolina. Marketing-conscious from the very start, Piper had this picture printed on postcards and sent them to prospective customers with the message, "Know your roofer."

ing launched their involvement with metal. But there were others who became composition roofers. Because composition roofing as we know it today was not really invented until the 1840s, metal roofs had a chance to claim the major share of specialists during the early years of the roofing industry. Many people, such as James T. Brodie and Richard Cumming of Philadelphia, were metal roofers who simply expanded their product line to include composition roofing when it became available. Others, such as Luther Barnes of Boston in 1845, entered the industry as composition roofers with no previous building experience.[23]

By the early 1840s, the roofing trade began to come into its own on the American East Coast. From there it moved inland; roofers appeared as a separate group in St. Louis in the early 1850s and Milwaukee in the late 1850s. At the same time, roofers arrived in San Francisco. This shift to the West mirrored the general movement of the frontier. From the West Coast, roofers began to move into the great western interior, entering Kansas City and Denver in the early 1870s.

The roofing industry of the 19th century was vital and dynamic. Opportunities in the field attracted a wide range of people from a variety of different occupations. James DeForest Candler, founder of the J. D. Candler Roofing Company of Detroit, is an example of someone who took a rather circuitous route to becoming a roofing contractor. The son of an English immigrant who had formed a house and ship joiner company in Detroit, Candler elected to enter the working world in his uncle's lithography shop in 1874. He worked there for about five years before he joined the tin, copper and sheet iron working firm of Frumveller and Snitgen, creating the firm of Frumveller, Snitgen and Candler.[24]

Candler's association with the iron workers ended quickly. The young man wanted to run his own business, and after learning the iron workers' trade, he sought work in bookkeeping. By 1882, he was ready to begin his own firm. Hc joined Charles E. Wrench to form the partnership of Wrench and Candler, "Manufacturers of all kinds of Galvanized Iron Cornices, Dormer Windows, Window Caps, Chimney Caps, etc."[25]

After two years as Wrench's partner, Candler severed the relationship. Like so many beginning building trade contractors, Candler sought out his brother, George, to form a new firm.[26] Using the skills J. D. Candler had developed working for others, the two brothers specialized in the manufacture of galvanized iron cornices. But within a short time, their firm expanded to include the application of iron work as well as slate, tin and iron roofing.

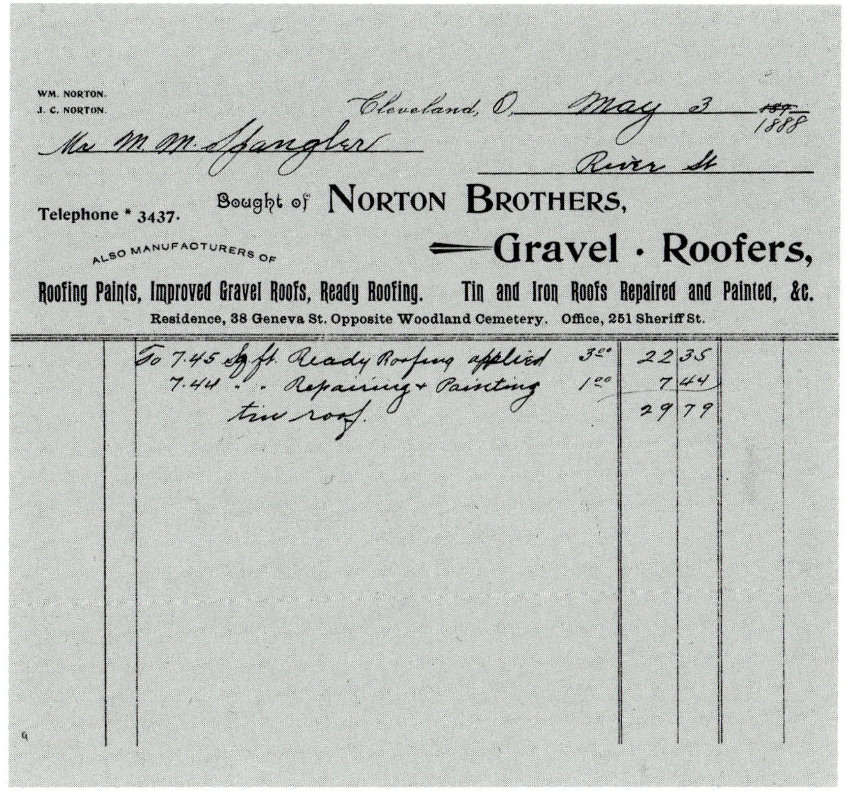

WM. NORTON.
J. C. NORTON.

Cleveland, O. May 3 1888

Mr. M. M. Spangler

River St

Telephone * 3437. Bought of NORTON BROTHERS,

Also Manufacturers of

Gravel · Roofers,

Roofing Paints, Improved Gravel Roofs, Ready Roofing. Tin and Iron Roofs Repaired and Painted, &c.

Residence, 38 Geneva St. Opposite Woodland Cemetery. Office, 251 Sheriff St.

To 7.45 Sq. ft. Ready Roofing applied	3.00	22	35
7.44 " " Repairing + Painting tin roof.	1.00	7	44
		29	79

An 1888 Norton Brothers invoice for roofing services. Application is billed at $3 a square foot; repair and painting is $1 a square foot.

Candler became a roofer because it was a logical extension of his manufacturing experience. It was this same opportunity that lured metal craftsmen to the roofs in the early 19th century. This type of integration also figured prominently in the growth of the most important 19th century contributions to the industry: composition roofing.

Johns-Manville Company's roofing manual, copyright 1903. The booklet describes the roof components and proper application procedures.

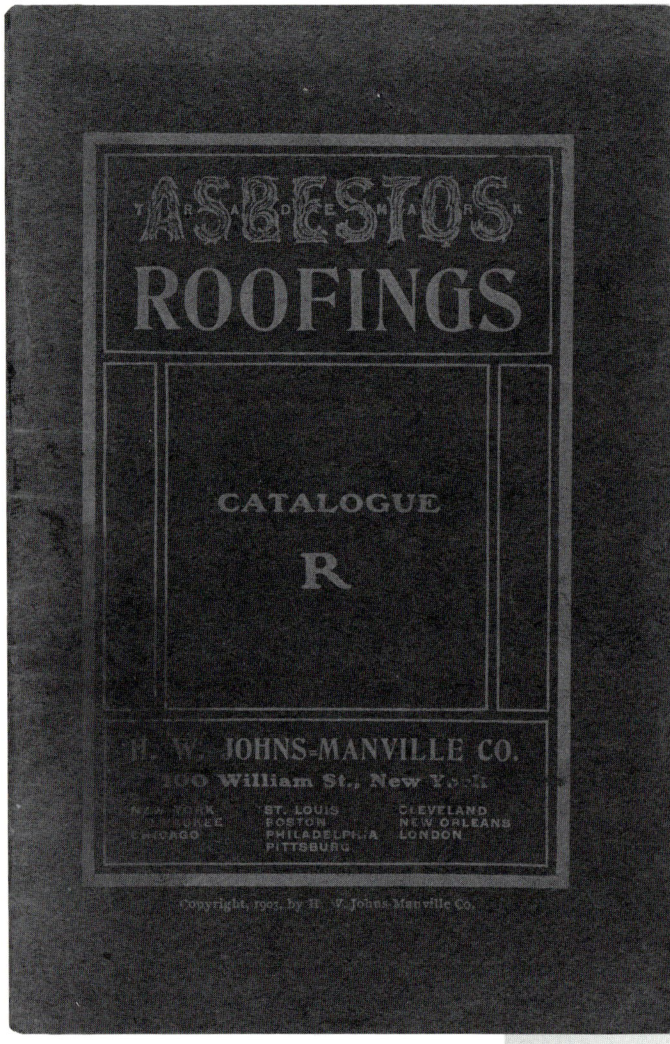

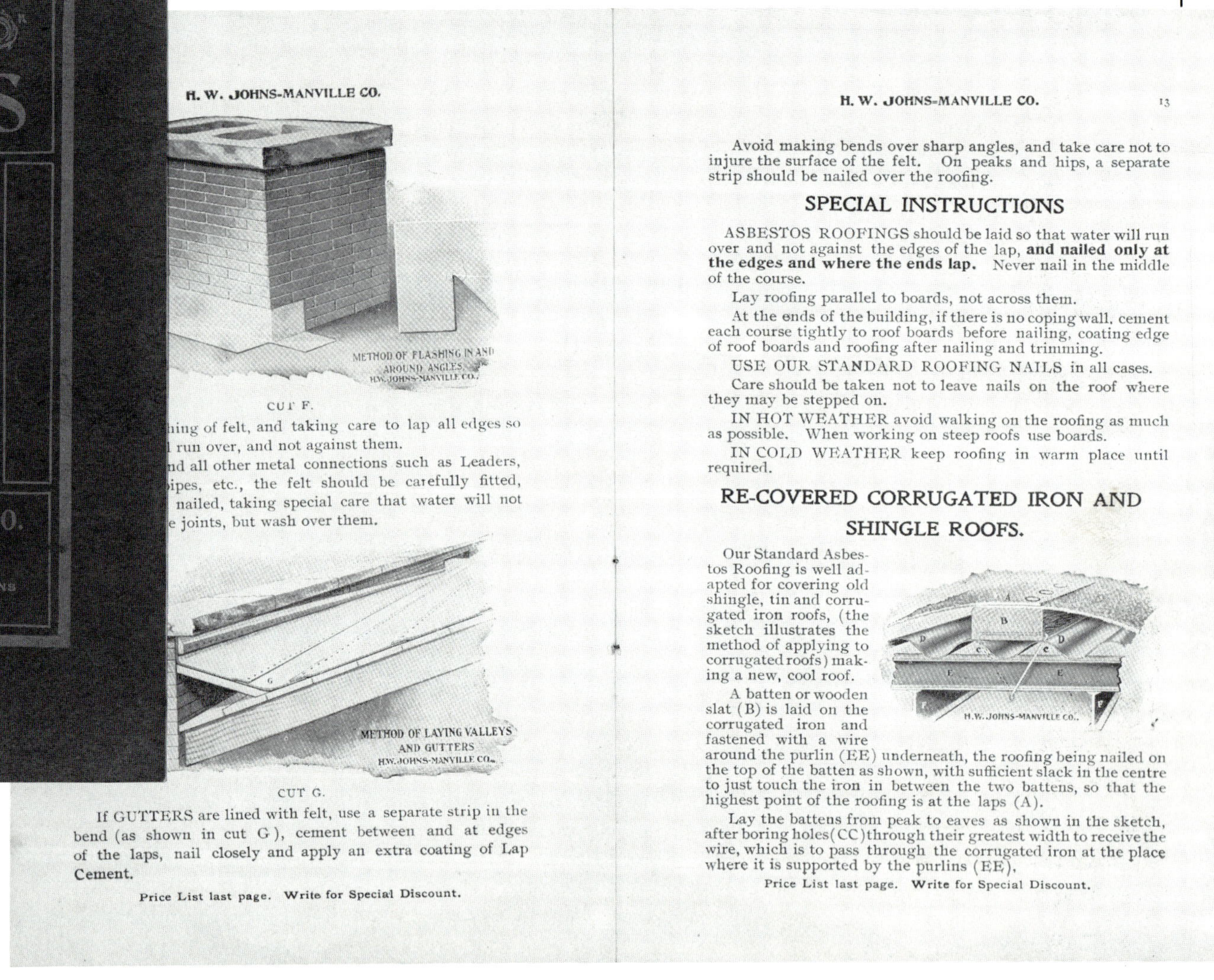

H. W. JOHNS-MANVILLE CO.

METHOD OF FLASHING IN AND AROUND ANGLES.
H.W. JOHNS-MANVILLE CO.

CUT F.

hing of felt, and taking care to lap all edges so l run over, and not against them.

nd all other metal connections such as Leaders, pipes, etc., the felt should be carefully fitted, nailed, taking special care that water will not e joints, but wash over them.

METHOD OF LAYING VALLEYS AND GUTTERS
H.W. JOHNS-MANVILLE CO.

CUT G.

If GUTTERS are lined with felt, use a separate strip in the bend (as shown in cut G), cement between and at edges of the laps, nail closely and apply an extra coating of Lap **Cement.**

Price List last page. Write for Special Discount.

H. W. JOHNS-MANVILLE CO. 13

Avoid making bends over sharp angles, and take care not to injure the surface of the felt. On peaks and hips, a separate strip should be nailed over the roofing.

SPECIAL INSTRUCTIONS

ASBESTOS ROOFINGS should be laid so that water will run over and not against the edges of the lap, **and nailed only at the edges and where the ends lap.** Never nail in the middle of the course.

Lay roofing parallel to boards, not across them.

At the ends of the building, if there is no coping wall, cement each course tightly to roof boards before nailing, coating edge of roof boards and roofing after nailing and trimming.

USE OUR STANDARD ROOFING NAILS in all cases.

Care should be taken not to leave nails on the roof where they may be stepped on.

IN HOT WEATHER avoid walking on the roofing as much as possible. When working on steep roofs use boards.

IN COLD WEATHER keep roofing in warm place until required.

RE-COVERED CORRUGATED IRON AND SHINGLE ROOFS.

Our Standard Asbestos Roofing is well adapted for covering old shingle, tin and corrugated iron roofs, (the sketch illustrates the method of applying to corrugated roofs) making a new, cool roof.

A batten or wooden slat (B) is laid on the corrugated iron and fastened with a wire around the purlin (EE) underneath, the roofing being nailed on the top of the batten as shown, with sufficient slack in the centre to just touch the iron in between the two battens, so that the highest point of the roofing is at the laps (A).

Lay the battens from peak to eaves as shown in the sketch, after boring holes(CC)through their greatest width to receive the wire, which is to pass through the corrugated iron at the place where it is supported by the purlins (EE).

Price List last page. Write for Special Discount.

CHAPTER 3

Necessity breeds invention: the birth of composition roofing

"Either I will find a way, or I will make one."

SIR PHILLIP SIDNEY

Today, composition roofing is one of the most popular and widely used systems in the roofing industry. Because built-up roofs are so ubiquitous, it is difficult to imagine a time when this type of covering was new or even revolutionary. In the mid-19th century, age-old systems such as wood shingle, slate, tile, or copper were still prominent in the roofing industry. However, in the span of a single generation, composition roofing burst onto the scene and assumed a leading place in the industry. The rather rapid spread of composition roofing techniques across the United States can be attributed in large measure to a small group of roofing pioneers: the Warren brothers of Cincinnati, Samuel Barrett of Chicago, and Michael Ehret of Philadelphia. What made these men so influential was not that they experimented with roofing compositions but that they were able to perfect techniques and forge business organizations that molded the history of American roofing.

To Samuel Warren must go the title of founder of the modern roofing contracting industry. He established one of the earliest roofing companies in the United States and pioneered composition roofing technology. Ironically, this man who occupies such a central role in the history of roofing had no desire to pursue a career in contract construction and abandoned the roofing business as soon as he could.

Warren was born in 1822, the son of a Massachusetts mechanic. Samuel and his three brothers had an unstable childhood. Their restless father had the family on the move constantly in his search for the proper town for his blacksmith and foundry shop. Between these various moves, Samuel received a good primary education. But his desire for higher learning was blocked by the Warren family's financial troubles. At the age of 15, Samuel Warren was apprenticed to a Vermont farmer. Under the terms of the agreement that removed the lad from his family, he was to work for six years on the farm. In return, he would receive room and board for the length of the apprenticeship and at its conclusion, a draft for $100 and a new suit of clothes.

The apprenticeship was negotiated between Samuel's father and Stowell Barnard, the Vermont farmer; the boy had nothing to say in the matter. Work at Barnard's farm was long and hard. Stowell Barnard had several daughters and perhaps he fancied being able to keep young Samuel on the farm as a member of the family even after the apprenticeship was completed.

Samuel would have none of it. After giving the apprenticeship a year's try and finding the prospect of life on the farm unappealing, he threw down his hoe and swore he would " . . . never dig another potatoe!" As Samuel packed his meager possessions and set off down the road, Farmer Barnard clenched his fist and cried, "You'll rue it . . . when you are twenty-one you'll think of that hundred dollars and suit of broadcloth clothes." But Samuel did not look back.[1]

He worked at various odd jobs. Like many rural youths without roots, he moved to New York City. Here he worked when able and got as much "enjoyment out of life as I could." Warren enjoyed the aggressive pace and uninhibited expression of opinions in this vital city. He even thought that he might "have some talent for public speaking."

Unfortunately, the slums of the city were also breeding grounds for disease and in 1840, he was afflicted with typhoid fever; he nearly died. His mother rushed to the city to nurse him. Samuel's long convalescence led him to reflect on his life; he decided he needed a goal and set an ambitious one. He resolved to make use of his delight in public speaking to become a lawyer. But to study law, he needed to earn enough money to receive a proper college education, and this proved to be a major obstacle.

At first he tried part-time classes while working as a postal clerk. In the predawn hours, he would pour over Latin and Greek texts, struggling to master the verse of Virgil. After several terms of such effort, Samuel was advised by his professors to leave college until he had "gone out into the world and made some money." A firm financial base would make advanced study easier. Samuel found it easier to take the first part of the professor's advice than the second. "It was easy enough to go out into the world; but to make money was another thing."[2]

Samuel Warren tried his hand at a wide range of jobs from schoolmaster to salesman, but none of them held the prospect of financial success. Later he recalled, "I got the name of being unsteady —a rolling stone that would gather no moss." He crisscrossed the eastern United States seeking opportunities that always seemed to vanish upon his approach. Because he saved his money for travel and lodging, he grew thin and gaunt from lack of proper food. He was down and out in Newark when his luck changed dramatically. As he walked along a street, his tired feet mechanically moving forward, he looked up to see a man applying a roof covering to a newly built house. As Warren watched the man at work, he noted that the roof he was applying seemed strikingly innovative, quite unlike the wood shingle roofs that were popular for the residences of the time. The roofer was laying out square sheets of heavy stock paper, coating them with pine tar, and then sprinkling sand on the surface. Warren had never seen a roof like it and, as he later remembered, "It struck me that I might do something with this."[3]

That evening Samuel Warren called on the Newark roofer he had met. The boastful fellow claimed that his system was taking the city by storm. He called the paper, pitch and sand application "Bacon's Patent Composition Roofing." After much discussion, Bacon and Warren struck a deal. Bacon claimed his process was unique and protected by federal patent. However, if Warren went to work for him, Bacon promised to teach the young man how to apply the roof and, in exchange for a royalty fee, grant Warren his patent rights for the Commonwealth of Massachusetts. Warren entered the roofing business. But the young Yankee's years on the road had taught him to be leery of the obvious good deal. He began work for Bacon, trying to learn the application process, but he also initiated some research into composition roofing systems.

The scholarly approach that Warren took toward composition roofing paid off. His inquiries at the United States Patent Office revealed that Bacon was neither the holder of a patent nor the inventor of a roofing system. Further correspondence revealed that similar roofs had been used by Boston builders for several years. Rather than confront Bacon with these facts, Warren shrewdly continued to learn the trade, biding his time before striking out on his own.

That opportunity came early in 1845. Nicholas Longworth, a wealthy Cincinnati merchant, was in Newark visiting relatives. He too had noticed the advantages of Newark's composition roofing. Longworth made further inquiries, which brought him into contact with Samuel Warren. Incredibly, Longworth offered Warren the initial capital necessary to begin a composition roofing company in Cincinnati. He further promised to use his considerable influence with Queen City merchants to develop clients for Warren. Samuel eagerly took the opportunity and headed west to Ohio, leaving the frustrated Bacon behind in Newark.

(Longworth's great-grandson, Nicholas Longworth III, would become one of the most influential men in the country. Elected to Congress at the turn of the century, he met and married President Teddy Roosevelt's daughter Alice in 1906. He eventually served as speaker of the House of Representatives and at the time of his death in 1931, was considered a potential candidate for president of the United States.)

Warren arrived in Cincinnati by steamboat. The levee area was a striking scene of excitement and commercial energy. Another visitor to the town described its active waterfront: " . . . colorfully painted three deck steamboats extended almost beyond the field of vision, most of them giving off smoke from two chimneys . . . the landing is covered with all sorts of bales, goods and barrels and is crowded with carts, porters, sailors, merchants, and departing and arriving travelers."[4] As young Warren walked down the gangplank, he found himself in a town given over to industry. Iron works, paper manufacturers, furniture factories, and other plants hummed with energy. Meat packing was the heart of the local economy. The long wooden sheds where hogs were slaughtered in the fall and winter were just one type of building in Cincinnati that could benefit from Warren's roofing process. The town had been wilderness fewer than 50 years before. By 1845, it was growing in frontier fashion: without any uniformity of architecture and without any restraint on individual initiative.

To properly take advantage of the opportunity that Longworth had provided, Samuel Warren felt he needed a business as well as a financial partner. He wrote to his family, then living in Glens Falls, New York, inviting his younger brother Cyrus to join in the roofing business. Cyrus was happy to leave the employment of his father's perpetually troubled plough shop and accept work in a growing town and a dynamic trade. After several months, a partnership was formed: S. M. & C. M. Warren.

With Longworth's contacts and the advantages of the composition roofing system, the Warren's business thrived. S. M. & C. M. Warren had done nothing to change the roofing business; the composition system that Samuel had learned in New Jersey enjoyed moderate popularity throughout the northeastern United States, and was being installed by other builders who specialized in roofing. But sand and pine tar left a great deal to be desired as roofing materials. Before a truly successful composition roof would emerge, experimentation was necessary.

The Warren brothers, while they set up business with Bacon's process, were more than willing to exercise their Yankee ingenuity to improve it as

A turn-of-the-century engraving illustrates the composition roofing process.

were many other would-be inventors. Many early composition roofing specifications read as though they were taken from an occult recipe book. In 1845, William Chase patented a process that coated roofs with a composition of mineral tar, crushed red sandstone and clay. Instead of felt or paper, some would-be roofers advocated flax, kaolin, cattle hair and even horse manure. For an adhesive, boiled fish oil, palm oil, and liquid glass were suggested at various times. As an external covering, sand was most often proposed, though exotic inventors did not hesitate to advocate marble dust, china clay, sawdust, or even ground oyster shells. Unfortunately for the imaginative men who experimented with composition roofs made of beef tallow, bullock's blood, or boiled rice, their efforts ultimately were abandoned.

The Warrens first revolutionized roofing in 1847 when they experimented with coal tar. Cincinnati, like many industrial cities, adopted gas street lamps in the 19th century. Gas for the lamps was provided by the Cincinnati Gas Light and Coke Company, which was established in 1843 to produce gas from coal and pipe it throughout the city.[5] An unfortunate residue from the conversion process was a sticky, black substance known as coal tar. The general policy of the infant gas industry was to dump this waste material into the nearest stream. This was not unusual, as the Ohio River served as the waste disposal system for other Cincinnati industries. Black globs of coal tar floated downstream with the emptyings of chamber pots and the offal and blood of the slaughterhouses. Samuel Warren had a better idea. He approached the Gas Light and Coke Company for permission to take all the waste tar off its hands and the utility accepted Warren's offer.

In one stroke, Warren greatly increased the profitability of his business. The coal tar worked as well as the pine tar then used by pioneer composition roofers. Warren received his coal tar free, while pine tar during the mid 19th century was increasing in demand and cost. The pine belt of the Southeast was pressed to supply materials for the printing industry, rubber product manufacturers, lamp oil makers, and paint manufacturers, in addition to the traditional naval stores market. The major user of pine products in Cincinnati was Procter & Gamble, which used resin in its manufacture of soap and candles.[6] Instead of competing with other industries for a diminishing resource, Warren secured a very low-cost, abundant substitute.

The Warren brothers next turned their attention to the materials that actually composed the roofing system. Instead of using sand as the unfortunate Mr. Bacon had taught him in Newark, Samuel Warren experimented with different types of gravel. Eventually, the Warrens settled upon a very fine gravel to serve as the outer covering of their roof. The gravel worked particularly well with sheets of felt fabric. The Warrens substituted rolls of paper or felt for the square sheets of ship sheathing paper that, with great difficulty, considerable waste, and unfortunate mess, had to be hand-dipped into barrels of tar. The felt rolls were saturated with tar and then passed through a ringer to press out the excess adhesive.

In a mere two years, the Warren brothers had established a successful roofing contracting business and pioneered the use of new materials that were to make composition systems the center of the roofing industry. Tar, gravel, and felt rolls became the dominant materials in roofing. The Warren brothers brought to roofing not merely ambition and good business sense, but a curiosity about the nature of substances and a willingness to experiment. Unfortunately, it was these same characteristics that led first Samuel, then Cyrus away from roofing and into other fields of endeavor.

When the firm of S. M. & C. M. Warren was founded, the articles of partnership allowed Samuel to begin studying law as soon as business conditions would permit. Samuel Warren entered roofing only to make enough money to become an attorney. As early as 1846, he reduced his workload with the roofing firm and enrolled in Harvard University's law school. After a year of study in Massachusetts, Samuel, perhaps to be closer to the business, returned to Cincinnati and enrolled in law school there. Eventually, Warren was admitted to the bar, but he never practiced law. Perhaps the young man's years on the road had imbued him with a bit of wanderlust. In 1849, he caught the gold fever and planned to head west to California with a party of overland pioneers. However, on the very eve of his departure for the gold fields, Warren was dissuaded from going. The influence of his spiritual mentors in the Church of the New Jerusalem was crucial to his decision.

Samuel had been raised a Congregationalist, but had become interested in the doctrines of the New Church through conversations with his younger brother, Herbert Marshall Warren. When he finally committed himself to the New Church, it was wholeheartedly. He devoted himself to the study of theology and began to prepare for a career in the ministry.

The Warren family eventually contributed five brothers and a cousin to the business. It was a good thing that their mother, Betsy, proved so prolific, because none of the Warren brothers were content to stay in one place very long. When Samuel abandoned active participation in the roofing company, Cyrus invited his younger brother Herbert to help manage the contracting business. Part of the partnership contract included a clause that would allow Cyrus to become a full-time student when conditions permitted and still retain his share of the business—the same arrangement Samuel had made. With this provision clearly understood, Cyrus also invited his brothers Ebenezer and John to join the firm.[7] Cyrus began to work on expanding the firm's business and improving its products.

The first step was to dispatch Herbert Warren to St. Louis to establish a branch of the contracting business. When Herbert arrived in St. Louis in 1849, that city was the leading metropolis of the West. Steamboat traffic and trans-Mississippi commerce had developed a healthy local economy and a growing construction industry. While Herbert introduced the Warren process to St. Louis, his brother John was given the task of managing the family's experimental plant in Buffalo, New York. Beginning in 1850, the Warrens tried to improve the adhesive and waterproofing qualities of coal tar by refining the waste substances they received from the gas works. Their Buffalo refinery was the first attempt made by an American to refine coal tar. It was an important step not only in the development of roofing, but in laying the foundation of an important branch of the American chemical industry.[8]

The investments required for refining spurred the Warrens to secure their supply of coal tar. During the 1850s, the family prudently secured long-term contracts to haul coal tar from the municipal gas works of New York and several other large cities. Initially, the gas companies paid the Warrens to take the tar off their hands. In the case of New York's North River Gas Works, the Warrens were paid 50 cents a barrel. The Gas Works had been prohibited from dumping the tar into the Hudson River because it was thought to be responsible for diminishing the local fish population. The company was faced with the alternative of dumping the tar into the ocean, a considerable haul, when the Warrens arrived on the scene. Later, as composition roofing and the chemical industry expanded, the gas works found itself in possession of a valuable by-product and began to charge for the coal tar. By 1886, the Warrens were paying $2 a barrel for the raw tar. Nonetheless, the Warrens' aggressive pursuit of coal tar supplies helped secure the family's leadership in an industry that was rapidly becoming populated with both roofing applicators and suppliers.[9]

By 1852, Cyrus Warren felt secure enough to exercise his option to withdraw from active involvement in roofing and to pursue his academic interests. He moved his young bride and family to Cambridge, Massachusetts and at the age of 28, the successful businessman entered Harvard University. His younger brother Ebenezer accompanied him. After several years in roofing, Ebenezer wanted a college education.

At Harvard, Cyrus enrolled as a student of Louis Agassiz, one of America's greatest natural scientists. The master had a unique test for all of his young acolytes. Cyrus was shut in a room with a microscope and a small fishbone. After an hour, Agassiz returned and asked the bewildered Warren what he could tell the scientist about the fish. When Cyrus could offer no more than a few basic comments, Agassiz delivered an introductory lecture to him on the powers of observation and the scientific process. It was a powerful lesson for Cyrus, who thereafter emphasized the importance of detailed observation in both his business and scientific pursuits.[10]

Cyrus Warren was graduated from Harvard in 1855 and was nominated by Louis Agassiz to be a member of Phi Beta Kappa. Those college years had only served to whet Cyrus' interest in science, particularly chemistry. After graduation, he went to Europe to study chemistry in the great German universities, first at Heidelberg and later at Freiberg, Munich and Berlin. The Germans were just moving into the forefront of coal tar experimentation.

Samuel Warren, the "father of the modern roofing contracting industry." Warren was a pioneer of composition roofing technology.

Among the most important by-products of coal tar research was the production of aniline dyes. Spurred by what he had seen in the German laboratories and by his own experiments, Cyrus Warren elected to pursue this line of research full time. In Boston, he built a modern chemical laboratory where he studied the oils and lyes that were by-products of coal tar distillation to make roofing pitch. A separate plant, the Warren Color Company, was built to make commercial aniline dyes from the by-products. The Warren works were among a handful of American chemical companies that were trying to compete with Europe by establishing a domestic coal tar dye industry. One such firm in Boston, George S. Page & Company, had been in business since 1861.[11] Unfortunately, the Warren Color Company, like other early chemical dye firms, was unable to brook European competition and was forced to suspend operations after two years.

This did not daunt Cyrus Warren. While occasionally teaching chemistry at Harvard or at the Massachusetts Institute of Technology, he continued to keep abreast of changes in the roofing business. With his brother, Ebenezer, who had also studied chemistry in college, Cyrus began to experiment with pure natural asphalt as a roofing compound. The richest source of natural asphalt in the world was Pitch Lake in Trinidad in the West Indies. The lake was a depression that was filled with oil bubbling up from the earth. With time's passage, the sun had burned off the light gas and oils, leaving a thick, sticky pitch behind. The deposit had been used as early as the 16th century, when Sir Walter Raleigh caulked the bottoms of his ships with it.[13] Similar tar deposits in California near Los Angeles and Santa Barbara had been used for roofing in the 1820s and perhaps since the colonial period.[14]

The Warrens first became interested in Trinidad asphalt in connection with the paving branch of their wide-ranging business. In 1865, Ebenezer Warren established a road paving company in Washington, D.C. He tried to use manufactured coal tar as paving asphalt, but met with only limited success. In 1876, he tried Trinidad asphalt and found it a superior material. Pennsylvania Avenue in Washington, D.C. was the first street he paved with this new compound, and this prominent success induced the Warrens to experiment further with it.[15]

Since 1849, businessmen had tried unsuccessfully to exploit Pitch Lake of Trinidad for its rich asphalt, but one after another failed. To ensure success, the brothers consolidated their efforts. Cyrus, Ebenezer, and even Samuel (who temporarily put aside his Bible) became directly involved in the endeavor. Their partner in Trinidad was

Amzi L. Barber, whose own paving company had branch offices in five American cities. The Warrens were willing to join with a competitor in the paving business because they had their eye on using Trinidad asphalt in roofing. Cyrus Warren had perfected a process for using petroleum tar as a flux for Trinidad asphalt. Petroleum tar was then a mere waste product from the distillation of oil. Like coal tar before it, petroleum tar was initially regarded by refineries as a nuisance they were glad to be rid of. Cyrus Warren took the petroleum tar and mixed it with the thick Trinidad asphalt, which in its natural form was solid enough to walk upon, and produced a fine, easy-to-mix roofing pitch.[16] Throughout the 1880s and 1890s, the popularity of Trinidad asphalt in both roofing and paving was an important part of the success of the various Warren companies.[17]

The Warrens eagerly pursued Trinidad asphalt as a roofing pitch because of the increasing competition for coal tar among the various branches of the infant American chemical industry. Waste products of urban municipal gas works were sought by more and more roofing contractors, and manufacturers of products as diverse as chemical fertilizers and creosote (to protect ráil ties and telegraph poles) were also willing to pay dearly for what previously had been given away. During the late 1870s and 1880s, the Warrens' contracts with local gas works began to lapse. Renewal of these agreements came only at terms considerably less favorable than they had enjoyed earlier. Tar refineries multiplied in America after the Civil War. This forced a drop in the price of refined tar; at the same time, the cost of crude coal tar was rising.[18] Although it is only natural that the Warrens sought a more profitable roofing material to market, it is also true that the brothers had no one to blame but themselves for their trouble. Not only did the Warrens introduce coal tar refining to the roofing industry, but their network of family and friends ensured that their systems would be introduced and adopted throughout the United States.

Composition roofing in Chicago, St. Louis, and Philadelphia was greatly stimulated by the Warren brothers. The great pioneer roofing firms of Ehret-Warren, Barrett Manufacturing Company, and M. W. Powell all owed their early growth to the influence of the Warren brothers. Certainly a factor in the Warren brothers' pervasive influence in the spread of composition roofing was the family's boldness and innovation. Yet, strange to say, it was the Church of the New Jerusalem that helped make the Warrens so influential, and composition roofing so popular.

The Church of the New Jerusalem had been an important part of the private lives of the Warren brothers since the 1840s. The New Church was founded by Emanuel Swedenborg, a Swedish scientist who in 1745 gave up the study of "worldly science" and devoted himself to the Bible. Swedenborg placed a great deal of emphasis on the dichotomy between the spiritual world, which was transmitted from God through the Bible, and the physical world of the senses. For Swedenborg and his followers, the Second Coming of Christ was a spiritual return to earth, which in fact had begun in June 1770. Because of its emphasis on spiritualism, the New Church (as Swedenborg's followers referred to it) addressed moral issues not from the perspective of the overt act, but of the individual's inner motive. The New Church grew slowly in the late 18th century, but it was able to put down roots in England, Sweden, and in America by the beginning of the 19th century. Massachusetts became the center of the New Church in America and it was there that the Warren brothers were converted to its doctrines.[19]

By the time the Warrens repaired to Cincinnati to found their roofing company, the New Church was well established in Ohio, and was in fact using that city as a base of missionary activity. Although Cyrus and Samuel Warren devoted themselves to their business in its first few years, they also participated in the New Church activities in town. The brothers' growing prosperity and their intellectual vitality brought them into contact with some of the leading figures in the New Church. The most important of those was the Reverend Benjamin F. Barrett, a highly intellectual minister who was rapidly becoming the most prominent New Church spokesman in the Middle West. In the spring of 1848, Barrett became the pastor of the Cincinnati Society and he may have lived with the Warren brothers for a time.[20] What is known is that the Warrens had a major influence on Barrett and vice versa. It was during the spring of 1848 that Samuel Warren began to consider a career in the New Church ministry.

Furthermore, the energetic Reverend Barrett became acquainted with the Warren composition roofing system. The New Church was not well enough established to support its ministers satisfactorily. Lecture tours between New Church societies were especially difficult to manage. A good business investment would allow the Reverend Barrett the surplus capital to devote more time to lecturing and writing. Discussions between the Warrens and Barrett led to a joint business venture. The Warrens would loan the minister the capital, tools, and know-how to establish a composition roofing firm. But so as not to allow competition with the firm of S. M. & C. M. Warren, it was agreed that Barrett's business would be based in Chicago.[21] This move not only expanded composition roofing to Chicago, but it led to the establishment of one of the industry's most important companies: Barrett Manufacturing.

Benjamin F. Barrett established his roofing business in Chicago in 1848. The minister did not personally devote himself to roofing at that time. He continued to spend most of his time in Cincinnati, where he remained the New Church pastor. Barrett may have had a foreman in Chicago supervising roofing operations, but it is doubtful that the business amounted to much in those first few years. However, in 1850, Barrett resigned his pastorate. He still lectured occasionally in Cincinnati, but he now had more time to develop his roofing business in Chicago.[22] That business was initially headquartered near the Chicago River in a two-story frame house. Barrett's men heated coal tar in an open kettle in the backyard.

The choice of Chicago as the site for Barrett's roofing business was not accidental. Although Cyrus Warren later remembered making the suggestion to Barrett (" . . . I would start in Chicago, which looks to me to be a likely town"),[23] it is unlikely that the well-traveled Benjamin Barrett needed Cyrus' advice. Chicago was second only to Cincinnati as a western center for the New Church. Barrett knew that by 1849, the society had a parish of more than 20 members in the city.[24] More important, the Chicago society was headed by Jonathan Y. Scammon, a young lawyer, who after 1847 was one of the leading builders in the Chicago area.[25] It seems that Barrett chose Chicago for his roofing business because of New Church contacts, which would have assured a steady flow of contracts. The fact that Chicago in 1848 was a boomtown only made the move more attractive. The city was swollen with immigrants from the German states, Ireland, Sweden, and the northeastern United States. In 1848, the Illinois and Michigan Canal opened, making Chicago the gateway from the Great Lakes to the Mississippi Valley. As the trade and population exploded, so too did construction. By 1855, more than 2,700 buildings were being built in Chicago during a single summer.[26]

Moses W. Powell, a strapping Pennsylvania farm boy who came to Chicago in 1850 at the age of 19, was an early employee of Benjamin Barrett. During the summer months, Powell worked as a roofing mechanic for him. But when autumn came, Barrett closed his roofing company to devote himself to New Church activities. Moses Powell, like the other Barrett employees, was forced to seek other employment during the winter. Some years he worked in Arkansas as a lumberjack, cutting trees and rafting them on the Red River; other winters were spent working on a Mississippi steamboat.[27] Powell excelled in his summer work, eventually becoming superintendent of

Benjamin Barrett's business.[28] When the minister sought to retire from active involvement in the roofing business, Moses Powell was the logical choice to succeed Barrett and take over the firm. But that move was blocked by the arrival in Chicago of Samuel E. Barrett in 1855.

In small businesses, family ties are frequently more significant than experience or merit. Samuel E. Barrett was a close relative (probably a brother) of the Reverend Barrett, but he also was a man who had accumulated considerable business experience in his brief 21 years. Born in Cambridgeport, Massachusetts in May 1834, Barrett received a public school education before entering into business, first in Boston and later in Milwaukee. Although he was roughly the same age as Moses Powell, Samuel Barrett was able to bring to the roofing business not only family ties, but management experience and a small amount of capital. So in 1855, Reverend Barrett, like Samuel Warren before him, turned his back on roofing, leaving management of the business to a family member.[29]

Samuel Barrett was what people of his day admiringly called a "hustler." He had a tremendous amount of energy, a strong personality, an independent nature and an excellent sense of organization. The business he took over from Benjamin Barrett was a modest contract construction firm that operated on a seasonal basis. Under Samuel's leadership, the Barrett firm immediately adopted a more aggressive stance. Whereas the Reverend Barrett had been content to purchase his roofing supplies from his colleagues, the Warren brothers, Samuel Barrett made plans to produce his own pitch. Chicago had its first gas works installed by 1850. Samuel Barrett, taking a page from the Warren brothers' own book, approached the gas works about its coal tar. The gas works was only too pleased to give all its tar to Barrett, and in fact, even paid him for hauling it off their property.[30] Barrett then distilled the coal tar, creating two oils. The lighter oil was used as a saturant for roofing felt; the heavier oil was employed as an adhesive to hold the felt to the roof.[31] Although Barrett did not produce his own felt, he did arrange for felt rolls to be produced to his specifications, which also greatly increased the efficiency of his operations.

Barrett's expansion into the production of roofing supplies was logical, but expensive. To capitalize such ventures he needed more money than was at his disposal, so he decided to bring partners into his firm. His first partner was William C. Dow, an important Chicago area roofing contractor. Dow had made his fortune in slate roofing and was rather hesitant to become involved in the newer composition roofing business. For two years, they cooperated under the name Barrett, Dow and Company. But in 1857, William Dow had enough of tar and paper and went back to slate roofing. Barrett then turned to William H. Arnold, another slate roofer. This proved a happier arrangement and the two remained partners for many years.[32]

Moses Powell remained in the Barretts' employ after Samuel took over the firm. Powell was by this time highly experienced not only in applying composition roofs, but also in managing and bidding contracts. In 1859, he decided to put this experience to the test and establish his own roofing business. He found a partner by the name of Nathaniel B. Mansfield and opened an office in the heart of the Chicago business district. The proprietorship advertised itself as "Powell and Mansfield—Manufacturers of Felt and Composition Roofs and Roofing Materials." However, a caption in their notice also proclaimed, "We furnish the roofings known as Barrett Roofing."[33] It is doubtful that Powell actually did manufacture his own roofing materials. What makes more sense is that his firm remained allied with Samuel Barrett through the latter's growing list of roofing supplies. Eventually the two men joined forces in one business: Barrett, Arnold and Powell. The combined firm became the premier roofing concern in Chicago.

The Civil War broke out in 1861 and the career of Samuel Barrett, like those of hundreds of other contractors, came to a sudden halt. Barrett enlisted in the Union Army in July 1861. He was mustered into an artillery unit as a first lieutenant. Within a matter of months, Barrett was promoted to captain and placed in charge of his unit.[34] The First Illinois Light Artillery saw extensive action at Fort Donelson, Shiloh, Vicksburg and Missionary Ridge. Barrett proved to be a determined foe, both of Confederate forces and of opponents on the general staff. In 1863, at the siege of Vicksburg, Barrett angrily resigned his commission because his battery had been overlooked for new ordnances. His bluff worked and he continued to serve in the Army of the Tennessee, reaching the rank of major by the end of the war.

Samuel Barrett returned from the Civil War to find his business struggling.[35] Although Chicago continued to grow during the Civil War, almost doubling in population, competition in the roofing industry was intense. There were scores of composition roofing applicators in the city, not all of whom were careful. Traditional shingle, slate, or tin roofers and contractors were outspoken in their citicism of composition roofing systems. One such critic noted, " . . . of the thousands in use in Chicago, we have yet to find the first man who is fully satisfied with them . . . they are continually getting out of order, and few of them long continue water-proof."[36] Discussions on how to best revive the trade of Barrett, Arnold and Powell in the face of competition revealed major differences between the interests of Samuel Barrett and Moses Powell. Barrett wanted to concentrate on roofing supplies and other coal tar by-products, while Powell believed the firm needed to expand its emphasis on application. The two men agreed to disagree and dissolved their partnership.[37]

The dissolution of Powell and Barrett in 1869 marked an important phase in the growth of the roofing industry. Specialization between the manufacturer of roofing materials and their application had begun. Moses Powell acquired new partners but his long-term goal was to operate his own business. This he was able to do by 1873, when he founded M. W. Powell and Company. Powell and Barrett remained on friendly terms and Powell made use of the Barrett name in his firm's advertisements. In the wake of the Chicago Fire of 1871, Barrett was also temporarily forced to take on a new partner. Edward A. Kimball, formerly a fire insurance executive, joined Barrett because he recognized the safety factor in composition roofing, as opposed to wood roofing. The Chicago Fire did result in a major building boom in the city, including the construction of a great many flat-roofed buildings. Barrett, Arnold and Kimball posted record profits in the year following the fire.

Throughout the 1870s and 1880s, Barrett's roofing supply business flourished. But like any successful enterprise, Barrett, Arnold and Kimball faced many challenges in maintaining that success. In 1872, an epidemic struck the city's horses. Also, a strange disease sickened and killed many of the draft animals upon which the 19th century businessmen of Chicago relied. Not only was Barrett forced to cut back on deliveries from his plant, but the very basis of his prosperity was threatened. His contract with the local gas works of Chicago gave him all the coal tar he needed, provided he made regular pickups of the black liquid. Many other roofing contractors would have liked to have secured that contract for themselves. To stave off that threat, Barrett assigned 24 men to haul the tar wagons by hand. For three weeks, Barrett's two-man-powered teams made the three-mile trip to the gas works and preserved the lucrative supply contract.

During one business recession, Barrett found himself with thousands of gallons of roofing pitch and few contractors to buy it. With both selling and dumping the noxious material out of the question, Barrett decided to find a new use for the substance. He learned from the newspapers that swarms of grasshoppers were threatening Midwestern grain crops. Grasshoppers had ravaged the Western frontier continually for several years; whole townships were devastated. Finally in October 1876, the governors of Minnesota, Kansas, Iowa, Missouri, and the Dakota Territory met in Omaha to consider the problem.[38] There were many suggestions about stopping the grasshoppers, from mass prayer meetings to the use of explosives. With such ideas being bandied about, Samuel Barrett did not feel embarrassed offering his solution to the plague. He called for sheets of iron to be laid in the path of an oncoming swarm. His excess coal tar would be spread on those plates. Once the plates were covered with grasshoppers, the compound could be ignited and the insects would be destroyed.

The cost of sheet iron and the use of fire were problems. Grasshopper plagues generally come coupled with drought. Nonetheless, farmers experimented with his approach and he was able to sell an extra 7,000 barrels of tar.[39]

Michael Ehret, who patented his slag roofing system in 1868. He had entered the carpentry business in 1860, when he was 15 years old.

It was this aggressive style that made Barrett's business ventures thrive. By 1889, his firm was successful enough to require reorganization. Operations had become too complex to be managed under a loose proprietorship. The newly formed S. E. Barrett Manufacturing Company took charge not only of the Barrett Illinois plants, but also the facilities of five other roofing manufacturers in the Midwest. With this reorganization, Barrett became a major force in the roofing industry. What had started as a minister's attempt to gain financial security had resulted 40 years later in a formidable manufacturing corporation.

One of the five companies absorbed was the Ehret-Warren Manufacturing Company. This company, with coal tar distilleries in St. Louis and Kansas City, was the Midwestern branch of the Warren-Ehret Company of Philadelphia. Warren-Ehret, like Barrett Manufacturing, was one of the pioneers of composition roofing. Warren-Ehret also owed its origins to the Warren brothers of Cincinnati and their connections to the Church of the New Jerusalem.

Like his brothers, Herbert Warren was a devout member of the New Church. When his brothers sent him to St. Louis, he became acquainted with the Reverend N. C. Burnham, a New Church minister and homeopathic physician. At that time, Burnham was tutoring Samuel Warren for a career in the New Church. Burnham was on intimate terms with the Warren family, and like Benjamin Barrett, he became interested in the roofing business.[40] The church called Burnham away from St. Louis before any formal arrangement could be made. But Samuel and Herbert Warren followed the Reverend Burnham to Philadelphia in 1853, and there formed "Warren and Burnham Composition Roofers." It is probable that Burnham, who was active in a movement to train young men for the Church, may have been given a share in the business in lieu of tuition.[41]

Once the Warren brothers were introduced to Philadelphia, they remained an important part of the roofing industry there for more than 30 years. Burnham remained associated with the company for only a few years. After securing a profit, he withdrew from the roofing business. By 1856, the Warren Philadelphia branch was known as the H. H. Warren Company. Ebenezer Warren was the brother who eventually assumed leadership of the Philadelphia branch. With the help of several partners (Howard Kick, John Lober, and Samuel Foster), he was able to lead the firm into the production of roofing materials and the application of roofs and road pavements. Ebenezer Warren used his profits from roofing to invest in real estate and art. From real estate, he realized another fortune that he invested in French paintings. Gradually he became more interested in collecting and the work of the Pennsylvania Academy of Fine Arts than he was in his coal tar distillery and roofing business.[42] In June 1883, he agreed to sell his share of the business to Michael Ehret.

Michael Ehret was a worthy successor to the Warren family. Like the Warrens and Barretts before him, Ehret was a man of bold action when it came to business. Unlike other composition roofing pioneers, Ehret was educated in the building trades, not in the classroom. His father was a carpenter. During the summer months, father and son worked side by side, building houses and sheds in Philadelphia. Any spare time the Ehrets had was used to attack blocks of wood with a maul and froe to produce shingles. In 1850, at the age of 16, Ehret became a carpenter and began to develop his own clientele.[43] Perhaps the years spent making wood shingles led him to seek an alternate covering for roofs. Whatever the cause, Michael Ehret entered 1860 as one of the pioneers of composition roofing.

Initially, his partner was another young man, Charles Volkner. With $1,000 borrowed from one of their relatives, the two youths entered the roofing business. Like other pioneer roofers, Ehret was actually able to collect a fee from the Philadelphia Gas Works for removing coal tar from its

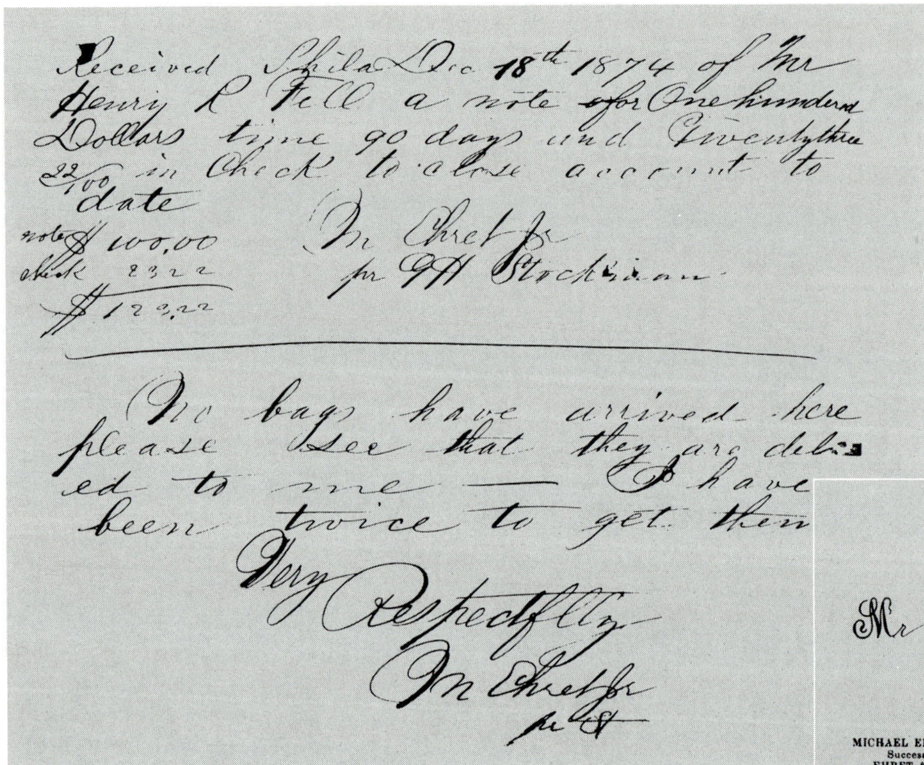

Received Phila Dec 18th 1874 of Mr
Henry R Fell a note ~~for~~ One hundred
Dollars time 90 days and Twentythree
22/100 in Check to close account to
date

note $100,00
Chk 23 22
$123,22

M Ehret Jr
per GH Stockman

No bags have arrived here
please see that they are deliv-
ed to me — I have
been twice to get them

Very Respectfully
M Ehret Jr
per G

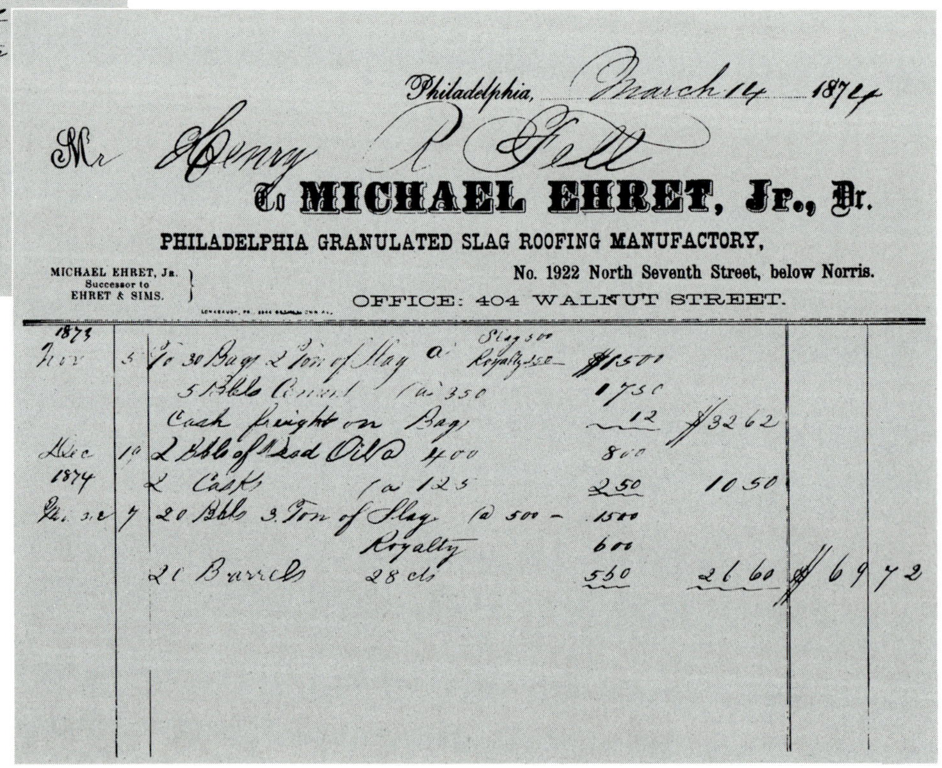

Philadelphia, March 14 1874

Mr Henry R Fell
To MICHAEL EHRET, Jr., Dr.

PHILADELPHIA GRANULATED SLAG ROOFING MANUFACTORY,

MICHAEL EHRET, Jr. Successor to EHRET & SIMS.

No. 1922 North Seventh Street, below Norris.

OFFICE: 404 WALNUT STREET.

1873 Nov	5	To 30 Bags 2 Ton of Slag @ Slag 500 Royalty 250	$1500		
		5 Bbls Cement @ 350	1750		
		Cash freight on Bags	12	$3262	
Dec	10	2 Bbls of Dead Oil @ 400	800		
1874		2 Casks @ 125	250	1050	
Mar	7	20 Bbls 3 Ton of Slag @ 500 –	1500		
		Royalty	600		
		20 Barrels 28 cts	560	2660	$6972

Michael Ehret grants Henry Fell the exclusive right to use Ehret's patented slag roofing system in Mercer County, New Jersey. The year is 1874 and the fee is $2 per material ton, making Mr. Fell one of the first licensed applicators of a manufacturer's roofing system.

grounds. The partners' work day began at 8 a.m. when they reported to their roofing job. They worked at laying roofs until 4 p.m., when they would take their wagon and team back to their warehouse. There they would spend several hours preparing pitch and felt for the next day's work. Ehret experimented with sand, gravel, and even seashells as a top coating before settling on slag.[44]

In 1868, Michael Ehret patented his slag (or cinder) roofing system. This was one of the first published composition roofing specifications. Ehret backed those specifications with a guarantee.[45] In time, he created a vertically integrated company. His operation grew to include slag-processing plants, felt-producing mills, and a coal tar processing plant. Not only did he manufacture all that was needed to produce a built-up roof, but he sold it commercially as well as applied it himself. Michael Ehret's 1883 decision to merge his operations with the remaining partners of the Warren company (Lober and Foster), further expanded the range of his firm. The newly formed business, Warren-Ehret Roofing Company, allowed Ehret to separate his application and retail interests from his manufacturing business. The company was capitalized at $30,000; $18,500 in "tools, fixtures and machinery, etc., and $11,500 in cash."[46] It had offered 300 shares at $100 per share, of which Michael Ehret bought 150 shares; Samuel Foster bought 75 shares, and John Lober bought 75 shares.

A general solicitor was retained to generate business for the new firm. He was to be paid a salary of $1,500 plus a 5 percent commission on all new sales and a 2.5 percent commission on "all work obtained from any of the present customers of either M. Ehret, Jr., or Warren Foster Co. who would naturaly [sic] come to this company with orders . . . "[47] Both the Ehret interests and the Warren interests were to supply the solicitor with a list of these potential clients. Business went very well and by the end of 1893, after only seven months of operation, the firm had received almost $50,000 in sales.[48]

In 1887, Michael Ehret, Jr. incorporated the Warren-Ehret Company in a move to consolidate all his manufacturing interests with those of the old Warren concerns. Ehret himself subscribed to 200 of the 1,000 shares of stock, as did George Elkins and George Widener (both partners with Ehret in another company), John Lober, and Harry S. Ehret.[49] The first order of business for the new company was to offer $90,000 for the purchase of Warren-Ehret Company Limited. At a special meeting, held the same day as Warren-Ehret's incorporation, Warren-Ehret Company Limited voted to accept the purchase offer. Although many of the same principals were

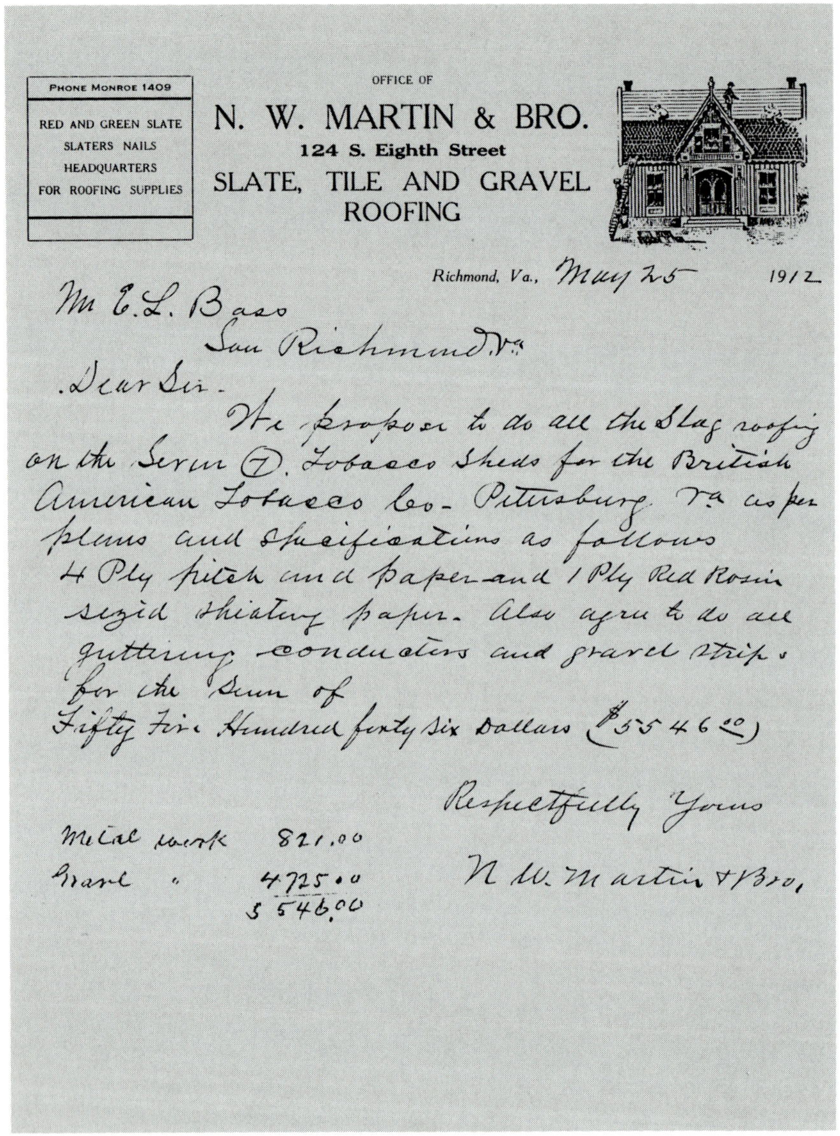
PHONE MONROE 1409

RED AND GREEN SLATE
SLATERS NAILS
HEADQUARTERS
FOR ROOFING SUPPLIES

OFFICE OF
N. W. MARTIN & BRO.
124 S. Eighth Street
SLATE, TILE AND GRAVEL
ROOFING

Richmond, Va., May 25 1912

Mr E. L. Bass
Sou Richmond Va

Dear Sir.
We propose to do all the Slag roofing on the Seven (7). Tobacco Sheds for the British American Tobacco Co.- Petersburg Va as per plans and specifications as follows
4 Ply pitch and paper and 1 Ply Red Rosin sized sheathing paper. Also agree to do all guttering conductors and gravel strips for the Sum of
Fifty Five Hundred forty six Dollars ($5546.00)

Respectfully Yours

N. W. Martin & Bro.

Metal work 821.00
Gravel " 4725.00
5546.00

N. W. Martin's 1912 bid to roof seven buildings for the British American Tobacco Company in Richmond for $5,546.

involved, the stockholders of Warren-Ehret Company Limited had increased their original $30,000 investment by $60,000.[50]

In 1888, Warren-Ehret reached $212,242 in sales. More than $150,000 of this was achieved through retail and factory sales of roofing components, while roofing application accounted for $41,000 and paving for $46,000. Despite the impressive sales figures, profits were very low because of the need to cut prices to meet some rather stiff competition. The 1888 profits were $2,678 or 1.25 percent of the total sales.[51] To raise capital, Ehret was forced to sell his plants in St. Louis and Kansas City to Samuel Barrett. The sale allowed Warren-Ehret to consolidate, but it marked the emergence of Barrett Manufacturing Company as a national force in the composition roofing business. If Michael Ehret was going to expand, it would be in the face of an equally strong and aggressive opponent.

By 1889, composition roofing was a well-established part of the contract construction industry. Because of the inventiveness of the Warren brothers, the network of the New Church, and the business acumen of Samuel Barrett and Michael Ehret, built-up roofs were in use throughout the United States. But the industry was young and the direction of its growth was unclear. Would the roofing industry be dominated by large national corporations such as Barrett Manufacturing and Warren-Ehret, or would the small "one-kettle" contractors inherit the future? The United States at the end of the 19th century was in the grip of rapid industrial and social change. For roofing contractors trying to adjust to great changes in their industry and in society, the future was anything but clear.

The proud Joseph A. Sanders, Buffalo's "live wire" roofer, poses with his family and company truck. Mr. Sanders started his business with $100 in 1919; he had been an apprentice sheet metal worker and roofer since the age of 13.

CHAPTER 4

Man and his machines: roofing in industrialized America

"Progress—the onward stride of God."

VICTOR HUGO

In December 1877, Moses Powell was an angry man. Not for the first time in recent months, he had received a letter from an outraged client. In spite of a recent reroofing job, the man's ceiling plaster was ruined by leaks. Sloppy work and furious clients were things Powell took pains to avoid. When they did occur, the employees responsible knew they would catch hell from "the Colonel." But what made Powell so angry in this case was that his job files showed no recent contract with the complaining client. In fact, the sloppy roofing job had been done by another contractor who used the name "Powell Roofing Company" to snare customers from M. W. Powell and Company. Even worse than the loss of clients was the effect the rival firm's sloppy techniques was having on Powell's reputation in particular and that of composition roofing in general. Powell immediately printed an advertisement disclaiming any association with "Powell Roofing Company."[1] After Colonel Powell calmed down, he may have reflected that the distasteful episode was not without significance. The pace of competition in the roofing business and in the economy at large was increasing. In the past, Moses Powell could have met personally with the few other contractors in Chicago to discuss common problems. But by the late 1870s, there were too many contractors scattered over the rapidly growing metropolitan area to maintain personal ties.

It was not only the growth in the number of firms that made the industry confusing for clients and competitive for roofing contractors, but the range of roofing systems. The United States Patent Office was deluged with new specifications for composition roofing alone. Everything from whale oil to sifted coal ashes to sulphate of zinc was proposed as an ingredient in built-up roofs.

But while composition roofers warred with one another over the vast number of specifications, paraffin paint manufacturers were trumpeting their product's superiority over the messy, smelly coal tar.[2] Michigan copper mines, faced with a downturn in demand for their ore, loudly proclaimed the superiority of copper roofs to composition systems. Of course, traditional roofs, such as sheet metal and slate, still had their supporters. All of these options and competing claims were even more confusing because American architecture, society, and business were in the midst of major change.

From 1870 to 1900, the American economy became industrialized. Before this time, most American factories and businesses directed their attention to meeting the needs of the nation's largely rural population. Businesses operating in such an economy were generally small proprietorships or partnerships exploiting local sources of supply and servicing local markets. But by the beginning of the 20th century, the economy was dominated by large national businesses servicing other businesses in the growing urban centers.[3] To fully appreciate the nature of change in the roofing industry during this period, it is necessary to appreciate the paradoxical phenomenon of the Industrial Revolution.

The Industrial Revolution started in England in the last half of the 18th century and reached the United States in the early part of the 19th century. Appearing first in the textile mills of both Lowell and Waltham, Massachusetts, it was characterized by an increasing presence of power-driven machinery and the disappearance of hand work. One of the earliest manifestations of American industrialization was the network of trans-continental railroads that spanned the country in the years after the Civil War. This network created the possibility of a national market for manufacturers. But in addition to opening a window of opportunity, railroads exposed manufacturers to the threat of competition from businessmen in other sections of the country.

Increased competition changed the nature of business. It marked the end of economic dominance by small proprietorships and partnerships, and the beginning of corporate America. Incorporation enabled a business to raise capital by selling stock to others, capital that was needed to buy the more expensive, efficient machinery being produced, and to build bigger factories.

Business and factory owners began trying to create bigger and better profits for their shareholders. Many companies diversified their products, and by doing so hoped to avoid being caught in the frequent economic slumps. Some companies diversified by buying other companies, thus becoming conglomerates. As the pace of consolidation quickened, cost-saving measures were adopted to increase profits.

Perhaps one of the most ingenious cost-saving methods was demonstrated by Andrew Carnegie, the steel magnate. Carnegie believed the best way to keep down the costs of Carnegie Steel Company and still increase production was to control all factors that contributed to the manufacture of his product. His steel was smelted from ore that was mined in the Lake Superior area. Traditionally, one company would mine the ore; another would ship it. By the time it was sold, however, the price had to cover the profit claimed by the mine owner and the shipper. Carnegie decided that his cost could be substantially reduced by cutting out the profit of the miner and the shipper. He bought his own mines; he shipped the ore on his own steamships; he produced the steel in his own mills. He effectively eliminated all middlemen from the process, and in so doing, created what is today called the vertically integrated company; that is, a company that controls all steps in material production from start to finish. Carnegie was so successful that when he sold the company to J. Pierpont Morgan after the turn of the century, it went for half a billion dollars, the first $500 million company in American history.

The corollary to the growth of industry was the concentration of population in urban areas. American cities increased fivefold in the period from the Civil War to 1900.[4] Most of this increase was from European immigration or from rural America. Cities were not prepared to handle such large influxes of people. Streets, lighting, transportation, sanitation, and primarily housing were strained to the breaking point. In 1893, more than half the people of New York lived in tenements. The 11th ward of that city rivaled Bombay in population density.[5] Conditions in the factories where the urban workers toiled were even worse. The working day varied from 10 to 12 hours; child labor was rampant; health and safety codes were nonexistent.

Men such as Andrew Carnegie, John D. Rockefeller, and J. Pierpont Morgan, who succeeded in creating national business systems, were decried by their workers as robber barons. The concentration of great wealth in the hands of a few individuals seemed unfair. By 1900, fewer than 2 percent of all industrial companies were producing 50 percent of all manufactured products. Farmers who previously had been the backbone of the American economy were frustrated by their loss of status and their growing dependence on the market economy. Organized workers railed against the factory owners, and the mass of newly arrived immigrant laborers undermined efforts to increase wages. America was divided on the question of which changes were good, which were bad, who was to blame, and what should be done.

A frustrated social critic cried out, " . . . cleavage of classes, cleavage of race, cleavage of faiths! An inextricable confusion. And the voice of democracy, crying aloud in the streets: 'out of this achieve brotherhood!'"[6]

At the time, the immense transformation that industrialization was bringing to society seemed chaotic. Social disorder in the form of strikes, riots, and government corruption was everywhere; so, too, was wealth and opportunity. Yet the lesson of industrialization that the great business trusts taught the companies they gobbled up and the consumers to whom they dictated was that strength lay in organization. It was not merely a matter of combining with others, but of organizing diverse elements into a working whole. For labor, this meant increasing interest in trade unions and socialism. Agriculture gave birth to the Farm Cooperative Movement. But it was among the urban middle class that the desire to organize—what one historian has called "the search for order"—was the strongest.[7]

The professions led the urge to organize and regulate. To cope with local government's inability to regulate the practices of medicine, physicians rallied around the previously neglected American Medical Association. In 1878, lawyers dedicated to improving standards in the legal profession formed the American Bar Association. Teachers, social scientists and architects all came together to define grounds for professionalism and opportunities for expansion. Businessmen formed local chambers of commerce or boards of trade. A growing number of organizations and associations that dealt with the problems of one particular business specialization were also popular.[8] The roofing industry felt the tug of these powerful national currents. While benefiting from the rise of industrialization, the industry also made its first moves to professionalize and regulate roofing contracting.

The first and most lasting example of this intent was the founding of the Gravel Roofers Protective Association. The word "protective" in the organization's title is revealing. The stability of roofing contracting was threatened by a wide range of forces, within and without the industry. Among the most important threats were the cutthroat competition among contractors, ineffective built-up roofing systems, poorly trained applicators, and perhaps most important, a growing restlessness among the labor force. The founders of the Association were Moses W. Powell and Samuel E. Barrett, named president and vice-president. Chicago's premier roofing contractor and the area's leading roofing supply manufacturer, those with the biggest stake in the industry, wanted to insure the health of roofing. Powell had personally suffered from the rivalry of unscrupu-

Moses W. Powell co-founded the Gravel Roofers Protective Association with Samuel Barrett in 1886.

lous and inept roofing contractors, while Barrett did not want the quality image of his composition roofing materials sullied by ineffective products. But the birth of the Association in Chicago was not merely the result of Powell and Barrett's natural desire to protect their businesses. Special historical circumstances combined to make Chicago the seed bed of early roofing association activity.

Chicago in the late 19th century was a microcosm of the nation. Its growth at the end of the 19th and the beginning of the 20th centuries was phenomenal, unchecked by natural disaster or depression. To house the thousands who thronged to the city for work in its factories, Chicago builders worked at a feverish pace. During the 1880s, developers caught "Flat Fever" and two- and three-story apartment buildings were put up by the thousands. In 1883 alone, 1,142 flat buildings were constructed.[9] This construction style greatly favored the city's composition roofing contractors, and led to an expansion of their ranks.

Because Chicago was an industrial boomtown, it was also the scene of the nation's most acute labor problems. Shortly after the Gravel Roofers Protective Association was formed, labor organizers and police clashed in the Haymarket Riots of 1886. Nightsticks flailed and bombs were thrown; crowds of passionate men surged from one altercation to another. The night of violence led to months of division between labor and management in the city. The Association allowed the roofing contractors to maintain a common front. Moses Powell, as Association president, represented the interests of gravel roofers in the Central Council of Chicago Builders. The purpose of the Central Council was to "promote the building interests of the city, [and] harmonize the different branches." But the impetus to cooperate was the belief that "labor threatened to throttle capital." The Central Council did not long endure, but its formation and Powell's participation in it did underscore the importance of industry cooperation.[10] The Gravel Roofers Protective Association had withstood its first test.

Three years later, in 1890, the Association was reformulated along national lines. The new National Association of Master Composition Roofers (also known as the National Association of

Master Gravel and Slag Roofers) was still dominated by Chicago contractors. The president, J. Wilkes Ford, and the secretary and treasurer were from the Windy City. Moses Powell was on the Board of Directors, but so were roofing contractors from Pennsylvania, New York, Michigan, and Indiana. The foundation had been laid for a truly national association.[11]

The infant roofing organization, however, was not equipped to deal with the full range of problems and opportunities that faced contractors at the turn of the century. The development of structural steel and continuous membrane covering went hand-in-hand, allowing industrial architects to cover great areas of floor space with flat roofs. The roofing of large industrial plants proved a boon to many contractors. Yet while new opportunities arose, old problems lingered. As urban areas became more dense, the threat of fire spreading from roof to roof became critical. Many cities had tried to deal with the threat by creating fire districts, where certain types of roofing materials were prohibited. But enforcement of such ordinances was hindered by an understaffed bureaucracy and uncertainty about what constituted safe roofing materials.

The danger of fire was compounded by builders' heavy reliance on wood construction materials. The balloon frame construction method allowed for the rapid building of homes and tenements. With the demand for housing high in industrializing America, contractors were loath to abandon the ease of wood construction for alleged safety reasons. Fires were common in American cities after the Civil War, just as they are today. But the severity of those blazes was much greater in cities made of wood. Chicago, like other towns, suffered from several major fires during this period. One of the worst was on October 7, 1871. It burned four blocks on the southwest side of the city. City fathers told themselves something had to be done about such blazes. But a catastrophe awaited them the next day.

Much has been written about Mrs. O'Leary's fabled cow and how the great Chicago Fire began that night of Sunday, October 8, 1871. Although the exact series of events that led to the conflagration are unclear, the fact is that the worst fire the country had ever seen resulted. From the sparks that set the O'Leary barn ablaze, a dry southwesterly wind quickly fanned the flames into a raging, ravenous behemoth, devouring everything in its path. By 2 a.m. Monday, the fire had consumed much of the present downtown and Loop area, and almost everything south of the Chicago River and east of its south branch.

Panic was rampant as the fire spread. People were running wildly about, trying to save what possessions they could. Those who had empty trunks and wagons to spare made small fortunes.[12]

Early Monday morning, the fire approached the residence of Lambert Tree, a Cook County Circuit Court judge. Judge Tree made a valiant attempt to save it. "I went up on the roof of my house," he related, "and ordered the servants to pass me up buckets of water as fast as they could, thinking that if I wet the roof thoroughly that would at least be a safeguard. In a few moments, however, I became convinced that no amount of water that I could command would save us."[13]

While on his roof, Judge Tree made these observations:

"The sparks and flaming felt were now flying as thickly on the North side as I had a short time before observed them in the South Division. The size of some of this burning material hurled through the air seems quite incredible. While on the roof of my house, a burning mass, which was fully as large as an ordinary bed-pillow, passed over my head. It fell upon the street, and upon descending, I had the curiosity to examine it, and found it to be a mass of matted hay. There were also pieces of burning felt, some of which I should say were fully a foot square, flying through the air, and dropping upon the roofs of houses and barns."[14]

Fed by the southwesterly wind and creating a breeze of its own, the fire advanced, an invincible force. It engulfed the great buildings of the city: the Court House (a portion of which had a "fire-proof" roof), the Post Office, and Farwell Hall. The city's fashionable hotels, the Tremont House and the Sherman House, roofed by Moses Powell with Barrett materials, burned quickly. The Sherman House staff had even placed men on the roof to put out the cinders that were landing there.[15] But no sooner would the fire engulf one building than the wind would blow it to its neighbor. In this fashion, the fire was able to jump the streets, the river and every barrier the citizens erected to contain it.

Many attempts were made to save the great houses of the town. At some homes, carpets were torn up, placed on the roof, and kept wet by water from the house's cisterns. But as one observer recalled, " . . . they might as well have tried to quench Vesuvius."[16]

The fire continued relentlessly on its path of destruction. The Gas Works failed Monday morning, leaving the city in darkness except for the eerie glow of the fire. Everywhere one looked, the fire had either done its work or was still burning. It was raining that night—not a rain that would aid in subduing the blaze, but a rain of sparks, embers and more fire.

An 1871 engraving of a Chicago city block in flames. The fire consumed 17,450 buildings in two days.

One small victory emerged amidst the destruction. By 8 a.m. Monday, the fire was continuing northward, toward Lincoln Park. The residence of Mahlon D. Ogden was in its path. The Ogden family had the good fortune to be out of town at the time, and the better fortune of having neighbors who could still put up a fight. With the aid of passersby, they beat back the fire on the roof of the Ogden house with wet blankets and aprons. The house was saved—the only house spared in the fire's path.[17]

By Tuesday morning, the blaze had finally died. It had burned 2,124 acres and consumed 17,450 buildings.[18] Save for the Ogden House, the destruction was complete. The sight that greeted townspeople the next morning inspired sorrow and wonder—so much had been destroyed by a power so great in so little time. One saw " . . . the destruction of the entire business portion of one of the greatest cities in the world. Every bank and insurance office, law offices, hotels, theaters, railroad depots, most of the churches and many of the principle residences of the city, a charred mass—property beyond estimate gone."[19]

The total monetary estimate of damages was $168 million.[20] Included in those losses, ironically, was the manufacturing plant of Samuel Barrett, producer of "fireproof" composition roofing.

Fires in the late 19th and early 20th centuries were brutal; total losses between 1860 and 1915 were more than $1.2 billion. In 1911 alone, 165,000 buildings burned. The magnitude of this destruction is evident if one envisions a street extending from Chicago to New York. Allowing each building a 65-foot front, this street would lose three miles of buildings a day to fire.[21] According to the National Board of Fire Underwriters, " . . . much of this conflagration loss can be traced to the flying brand hazard of shingle roofs."[22]

As one of the leading forms of roof covering in the 19th century, wooden shingles continued to wreak untold damage on American cities well into the 20th century. In May 1901, Jacksonville, Florida experienced a fire that started in its outskirts and proceeded through the main residential section to the business district. Shingle roofs ignited blocks of homes in advance of the fire because of the great heat. Total damage caused by this fire was more than $10 million.[23] Officials in Chelsea, Massachusetts found shingle roofs to have been a major contributor to an April 1908 fire that claimed 3,500 buildings, 18 lives and 275 acres.[24] The rapid spread of fires in Bangor, Maine; Houston, Texas; and Salem, Massachusetts were all attributed to wood shingles.

Augusta, Georgia, experienced a $4.5 million fire in March 1916. It claimed 560 buildings and 1,600 acres. "The spread of the fire . . . was due entirely to the sparks igniting shingle roofs on out-houses and dwellings . . . " according to Fire Chief Frank Reynolds. He continued, " . . . shingle roofs . . . ignited from a shower of sparks when the wind changed slightly towards the southeast."[25]

Attempts to save buildings by protecting their roofs and the general concern about roof flammability was summed up in an adage of the day: "If the roof can be saved . . . all is safe."[26] The destruction wrought by the Chicago Fire amply demonstrated that once the roof was lost, all was lost.

Turn-of-the-century efforts to provide fireproof roof coverings centered on the elimination of wood shingle roofs, and the enhancement of fire-repellent qualities of composition roofs. The solution hit upon was to subject each type of roof covering to a series of rigorous tests, and then place each in a certain class according to its ability to resist fire. Customers would have a good idea of the fire-resistant qualities of whatever roof they were considering installing.

The job of testing roofing materials went to Underwriters Laboratories, with the classifications officially being promulgated by the National Board of Fire Underwriters. The materials were subjected to flame tests that simulated an interior flame burning on the underside of a roof and a flame licking at the exterior of the roof. Burning brands were placed on the roofs to test their resistance to that method of ignition. Roofs were tested for the amount of heat that would pass through them on the theory that if too much heat was conducted, materials on the other side could ignite. The quality of the materials used to construct the roof, their performance, and the roof's ability to withstand the weather were also tested.[27]

Based on the results of these tests, the National Board of Fire Underwriters designated three general catagories of roof coverings in 1916. These were divided into classes. The standard (fireproof) roof included class A roofs and class B roofs. Fire-retarding roofs included those in classes C through F, while flammable roofs were classes G and H.[28]

Class A roofs provided the highest quality fire protection. These roofs were not easily set afire and would not maintain or communicate a flame. No gasses or flying brands were given off by these roofs, and they could effectively blanket an interior fire. They were considered the most durable type of roof, needing little repair. Class B roofs provided a relatively high degree of quality, but not the highest, and they would emit few if any flying brands.[29]

The qualities of the fire-retardant roofs deteriorated rapidly. Class C roofs afforded a moderate degree of fire protection. They did not generate many flying brands and could still act as an effective blanket, thereby smothering interior fires. Still considered durable roofs, they required more frequent repair than class A and B roofs. Class F roofs provided virtually no fire protection. They were easily ignited and communicated flame readily. Presenting a severe flying brand and spark hazard, they offered no blanketing ability for interior fires and required frequent repairs or replacement.[30]

Material manufacturers saw a valuable sales tool in their ability to put to rest people's fear of fire entering their homes through the roof. They clamored to have their various products tested and rated.

The roofing materials of Richardson/Flintkote, a company that had been in business since 1857, were rated either A or C. The company's class A 20-year roofs over wood decks, for instance, included one that offered a single layer of 30-pound felt and three layers of 15-pound felt, held together with four asphalt moppings and covered with slag or gravel. Its class C roofs were usually 10-year roofs, one of which had a single layer of 30-pound felt and two layers of 15-pound felt, three moppings and no slag or gravel cover.[31]

Johns-Manville, a major manufacturer since 1901, produced class A, general-purpose 20-year roofs over wood decks, which consisted of either one 60-pound layer of felt and three 15-pound layers, joined by four moppings, or five 15-pound layers of felt. The former had a smooth surface, while the latter had a gravel/slag covering. Its general-purpose roofs retained a class A rating, through the 10-year roofs that had four 15-pound layers of felt adhered with three moppings.[32]

Composition roofs were not the only ones to receive class A ratings. Tin was a fireproof roof covering that had been highly regarded from the time of its introduction into the United States. Tin roofs did not allow a fire to penetrate either side.

Tin emitted no flying brands and, when seamed properly, held together well even under the intense heat of a fire, thereby providing good blanketing capability.

By 1912, the National Board of Fire Underwriters considered metal, slate, tile and approved composition roofs the best for fireproof construction, yet each of them still had problems. Metal roofs, while being spark-resistant, might buckle under the heat. Slate and tile roofs could crack and pop nails, allowing the material to fall off the structure. The tar in composition roofs could melt under the heat of a fire, causing the roof to slide.[33]

Shingle roofs were considered totally unacceptable. To remove them from the market once and for all, the National Board of Fire Underwriters sponsored a 1916 booklet that called for the elimination of wood shingles as a building material. Augusta Fire Chief Reynolds summarized the argument:

"Personally, I have always opposed shingle roofs for the reason that they are conflagration breeders, and during a high wind and low temperature we have had as many as seven fires in two hours. Twenty-nine shingle-roof fires occurred during the Riverside Compress fire of November 3, 1915. This is conclusive that shingle roofs are a menace to any city, hence the covering with shingle roofs should not be permitted within the city limits. Only fireproof roofings should be permitted in erecting new buildings or in making repairs on old roofs."[34]

This campaign against wood shingles gave birth to the asphalt shingle market. An outgrowth of the composition roofing industry, asphalt shingles were made from felt impregnated with asphalt. Typically covered with granulated minerals, they were not susceptible to sparks and were much more resistant to heat than their earlier counterparts. Their popularity received a big push when the National Board of Fire Underwriters recommended them for use in place of wood shingles.

How successful were these early 20th century efforts to reduce roof flammability? A review of the premiums charged by insurance companies for each type of roof is revealing. In 1903, the standard roof, against which other materials were judged, was metal—most likely tin. It was charged out at the basic insurance rate, whatever that was for the specific location and company. Approved class A composition roofs were assessed at an additional cent, and deemed nearly as safe as metal. Slate roofs were assessed an additional 2 cents, largely because of their tendency to crack under intense heat. The underwriters were afraid that the falling pieces could hit firemen on the head and allow drafts through the roof that would fan the fire. Surprisingly, insurers were willing to

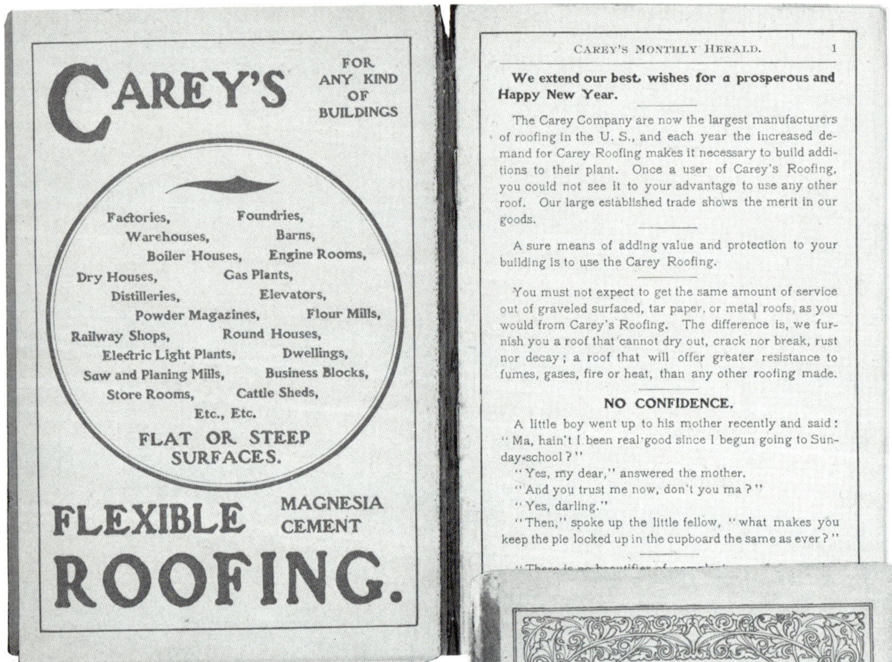

CAREY'S MONTHLY HERALD. 1

We extend our best wishes for a prosperous and Happy New Year.

The Carey Company are now the largest manufacturers of roofing in the U. S., and each year the increased demand for Carey Roofing makes it necessary to build additions to their plant. Once a user of Carey's Roofing, you could not see it to your advantage to use any other roof. Our large established trade shows the merit in our goods.

A sure means of adding value and protection to your building is to use the Carey Roofing.

You must not expect to get the same amount of service out of graveled surfaced, tar paper, or metal roofs, as you would from Carey's Roofing. The difference is, we furnish you a roof that cannot dry out, crack nor break, rust nor decay; a roof that will offer greater resistance to fumes, gases, fire or heat, than any other roofing made.

NO CONFIDENCE.

A little boy went up to his mother recently and said: "Ma, hain't I been real good since I begun going to Sunday-school?"

"Yes, my dear," answered the mother.

"And you trust me now, don't you ma?"

"Yes, darling."

"Then," spoke up the little fellow, "what makes you keep the pie locked up in the cupboard the same as ever?"

Carey's Monthly Herald, *"published for a minute of your time," combines The Philip Carey Manufacturing Company's (now Celotex's) advertisements with home-spun wisdom and notable observations, such as this from a 1904 booklet: "Within eight months, we have received enough emigrants from Europe to re-people Ireland."*

protect buildings covered with wood shingles, but they charged an additional premium of 15 cents, an expensive testimony to the system's flammability.[35] In actual practice, the percentage of total fires originating from exposure to, or sparks from, another fire dropped from an 1885-1890 five-year average of 42.93 percent to 33.12 percent in 1903, a decrease of 9.81 percent.[36]

Although tin's use as a roofing material continued to be extensive, the excitement generated by the development of the composition roof in the last half of the 19th century shoved it and other metals into the background. Tin manufacturers did not enjoy this position; an advertising battle ensued. Manufacturers of each type of material took careful aim at the other.

The Barrett Manufacturing Company fired some of the first salvos. In 1853, the company declared its new composition roofing material to be "an excellent protection against both fire and water, and very cheap compared with metallic roofs . . . "[37] Barrett further claimed that its roof would withstand a fire five times longer than would zinc, tin or galvanized iron. Johns-Manville picked up the battle cry in 1872 when its literature included a testimonial that declared the roof of the Tribune Building in New York, which had been covered with tin, now covered with Manville asbestos roofing and "in perfect condition." It seems that the building owner felt the tin roof failed because of the stress that expansion and contraction placed on the soldered seams.[38]

Perhaps the most effective charge against tin came from a composition roofing manufacturer. In 1872, Congers fired off a litany of charges against tin and metal roofs. They were susceptible to blowing off, Congers contended, and likely to rupture because of expansion and contraction. They were further vulnerable to rust, and even more rapid deterioration by the new industrial acids polluting the air of industrialized America. Congers further stated that " . . . the noise of falling rain and hail confuses the house from garrett to cellar . . . "[39]

Early 20th century advertising revealed a changing tide as the tin roofing interests seized upon the results of a 1911 Fire Underwriters test. In that experiment, a composition roof ignited when exposed to a flame and a tin roof did not. As a result, tin manufacturers reasoned that "sheet-metal roofers have a useful field of effort in the revision of obsolete building codes in those cities that still permit the use of roofs composed of flammable materials."[40] The tin roofing industry also jumped on the results of an Underwriters Laboratories test that rated the fire-retardant qualities of "good slag and good gravel roofs" in a class with inferior roofing materials.[41]

A rather spirited polemic against tar and gravel roofs was published by the Edwards Manufacturing Company, a tin roof manufacturer, in 1916:

"These composition or 'prepared' roofings all have a base of tar or asphalt on paper, felt or burlap body. Can you imagine anything that would catch fire easier? We are ready to offer $100.00 for any sample of composite roofing that will not ignite when a lit match is held under it. Try it yourself and see. Then think of having a roof on your house or barn that invites destruction like that. Think of exposing your family, your livestock and your machinery to such constant danger. You

An example of "Barrett-approved roofers" in 1916. Barrett placed a great deal of emphasis on good workmanship.

don't save anything by using composition roofing, for it only lasts three or four years, and then you are put to the expense of buying new roofing and the cost of time and labor in putting it on. In the summer the tar and pitch boils in the sun and clogs the gutters and rain pipes, a splendid breeding place for germs. It dries and cracks and soon becomes mere dried pulp with no protection whatever against fire, water or weather."[42]

The best defense against the vocal criticism of composition roofing was to improve the overall quality of materials being applied. By 1911, there were more than 300 companies manufacturing prepared roofing. There was no standard by which to judge the quality of their various products. Taking a cue from the contractors, roofing supply manufacturers united in an association.

The Prepared Roofing Manufacturers Association was established in 1911 to professionalize the production of roofing materials. The Association, which in 1926 changed its name to the Asphalt Shingle & Roofing Association, was able to use peer pressure and modest education programs to gradually improve the quality of materials with which contractors were supplied.[43]

Despite being eclipsed by the development of the composition roofing industry, the technology of tinplate manufacturing grew throughout the 19th century. Improvements included the introduction of sulphuric acid in 1829, and in the middle of the century, the use of rollers to transport the plates through the grease and tin pots, and spring-loaded rollers to adjust the thickness of tin coats to be applied. Bessemer steel plates were introduced in 1880. The industry was moving ahead so fast that by 1893 there were even experiments to produce an electro-tinplate process.[44]

Regardless of its growing popularity as a roof covering, most of the tin being used in the United States was being imported from Great Britain. No tinplate had been produced in America before 1872, although some thought had been given to it. In 1868, for instance, the Cambria Iron Company sent a representative to England to learn the tinplate trade, but little came of that venture. By 1872, other companies had followed suit, culminating in tinplate works at Leech, Michigan; Demmler, Pennsylvania; and Wellsville, Ohio. Business for these operations never did develop, however, and they soon resorted to the infrequent manufacture of terneplate.[45]

Tinplate production in the United States finally got a big boost in 1890 when, emulating the British taxes of 1703 and 1706, the McKinley Tariff fixed a duty of 2.2 cents a pound on the tinplate imported to this country. The import duty finally gave American manufacturers the edge they needed to make tinplate production a profitable and practical venture.[46]

The tinplate tariff was a very emotional issue. It was also closely related to the political fortunes of Ohio congressman and future president William McKinley. Many Americans felt that the industry had not been developed in their country simply because manufacturers could not master the process. Continuing to believe that it could not be mastered, contractors were angered because all of their imported tinplate would now cost even more.

William McKinley had sponsored the bill when he was a member of the House of Representatives. It was signed into law in the year that McKinley was up for re-election. With many Americans upset with the tariff, and with foreign suppliers irate and threatening retaliation, McKinley lost his bid for re-election. When he ran for governor of Ohio in 1891, the tariff issue continued to hound him. Defending himself, McKinley asserted that tinplating was not "an occult art, masterable only by a Welshman."[47] McKinley was vindicated when he won the governorship, re-election in 1893 and the United States presidency in 1896.

A St. Louis can manufacturing firm, the Norton Brothers, was the first large-scale producer of tinplate in the United States. Norton imported tinning equipment from England and set about making tinplate for its own use.[48] The firm was followed by the N. G. Taylor Company, a Philadelphia company that had been producing terneplate since the 1830s.[49]

When the McKinley tariff was adopted in 1890, the Welsh had a monopoly on the American tin market. This was not to last, however, because in 1892 there were 20 tinplate works in the United States, with 10 more under construction. In 1894, there were 56 works in operation. American production grew rapidly, and by 1896, production surpassed British imports. By 1911, the American tin market was supplied entirely with domestically produced tin. The growth of the American tinplate industry can be demonstrated by the amount of steel that was used for plating. In 1891, it was less than 1,000 tons; in 1914, it was about 900,000 tons.[50]

A variation of tinplating was terneplating, another popular type of roof covering. Terneplates were ironplates coated with a tin and lead combination and first developed in the pre-industrial portion of the 19th century.

An 1858 British patent is often cited as the first reference to terneplating, but an American patent pre-dates that by 20 years. On September 29, 1831, Joseph M. Truman of Philadelphia was granted a patent for increasing tin's durability by coating it with lead. According to Truman: "The purpose which I have in view is to preserve the iron more effectually from rust, and consequently to render the plates much more durable than heretofore, when employed for the covering of roofs, the making of pipes, gutters, cisterns, bathing tubs, and various other articles."[51]

Early commercial production of terneplate in the United States was undertaken by N. G. Taylor of Philadelphia, using imported tinplate.[52] As terneplate came to be known as a roofing material in the latter part of the 19th century, it was preferred to tinplate. It was thought to have a heavier coating and thus be better able to withstand the weather. The fact that it was cheaper than tin did nothing to hurt its popularity with builders.

Although terneplate manufacture in the United States dates back to Truman's patent, it was never produced in large quantities until the American tinplate industry was born. At that time the lead/tin coating on the steel became a mixture of the two metals instead of a separate coat for each. In 1892, terneplate production exceeded 9 million pounds compared to 4.5 million of tinplate. The two-to-one ratio is attributed to the less exacting process of terne production and the fact that there was a larger demand for domestically produced terneplate for roofing than for tin.[53]

To the delight of the American terneplate manufacturers, their product was selected to cover the buildings of the 1893 Columbian Exposition, the Chicago World's Fair. The buildings at the Fair were designed under the coordination of a great architect, Daniel Burnham. Under his direction, each building at the Fair was designed by a different architect. Because it was the only style equally well known by all architects, and because it would promote a visual unity among all buildings on the grounds, the classical style was selected as the Fair's architectural theme.

As a result of the homogeneity of all the buildings, their placement in relation to each other and the type of landscaping used to present them, the Fair was the catalyst that ushered in the era of modern American urban planning. It was a widely visited and studied exposition, and terneplated roofs were there to bask in the close inspection.[54]

Galvanized iron, aluminum and copper were all metals that enjoyed a degree of success as roof coverings in the late 19th and early 20th centuries. The use of galvanized iron as a roofing material can be traced to about 1840 in New York City. But many contractors who used galvanized iron preferred to have it tinned beforehand. Ironically, this tinning had been suggested by I. A. Sorel, the Frenchman responsible for creating the galvanizing process. He theorized that a coating of tin would greatly enhance the appearance of the metal. A more practical advantage was stressed in an 1852 trade catalog. The roofs' durability would be enhanced by the tin coating. Regardless of the advantages of the new material, it apparently did not become popular, because it disappeared from builders' catalogs shortly thereafter.[55]

The practice of corrugating galvanized iron gave it a second chance. Corrugation provided the otherwise flimsy sheet of iron with great strength, to the point where it could be directly applied to a building's rafters without a deck. It was particularly well-suited for prefabricated buildings.

When gold fever gripped California in 1849, corrugated, galvanized roofing enjoyed great popularity. California's population increased a thousandfold overnight, from approximately 14,000 in 1848 to just less than 100,000 by the start of 1850. By 1860, gold diggers from throughout the Americas swelled the Golden State's popula-

The Dayton, Ohio shop of Wetzel and Schriber (now Schriber Roofing Company), founded in 1907. Hyman Schriber was a coppersmith from Latvia before coming to America.

tion to 380,000.[56] In an effort to house all the new arrivals, contractors constructed prefabricated buildings, many with corrugated galvanized roofs.[57]

The use of galvanized iron increased and was a popular roofing material at the turn of the century, but there were problems with it that kept it from becoming a major force in the industry. The principal difficulty was the tendency for the zinc to flake off when the material was bent, leaving the iron exposed to rust.

Copper made somewhat of a resurgence as a roof covering in the last half of the 19th century. This was because of new deposits found in the Lake Superior region of Michigan's Upper Peninsula. In fact, these deposits permitted domestic copper production to double between 1860 and 1873. The discovery of these deposits was coupled with a technological breakthrough: the electrolytic process of copper refining. By 1892, builders believed that copper, along with terneplate and tinplate, was poised to lay claim to a large portion of the roofing market.[58] But copper was not destined to achieve such a position. Problems inherent in the material itself, its susceptibility to expansion and contraction, and the corresponding damage to seams, as well as the chemical reactions copper causes when in contact with any other metal, inhibited its popularity.

Unlike copper, galvanized iron, tinplate and terneplate, all of which had their origins in the first half of the 19th century or earlier, aluminum was a product of late 19th century technology. Two processes for refining aluminum were developed: one by E. H. and A. H. Cowles in 1885 and the other by C. M. Hall in 1886. As a result of their efforts, aluminum production in the United States grew from about 80 pounds in 1883 to more than 7.5 million pounds in 1893.[59]

The Philadelphia City Hall, built in 1891, was one of the first buildings roofed with aluminum. It had been specified as a coating for the iron and steel sheets that would cover the building.[60] Unlike zinc, however, aluminum did not make its mark as a coating for other roof coverings. Rather, it became recognized as a roof covering on its own merit. But there was a problem. It was difficult to solder the seams of the new metal to guard against leakage. The turn-of-the-century solution was not as much a solution as an avoidance of the matter. Aluminum was simply used on roofs pitched steeply enough to make the soldering of the seams unnecessary, as water never ponded on them.[61]

The roofing industry, along with the construction industry in general, had been growing rapidly in the late 18th century. But with that growth came a new problem: quality control. Actually the problem was not new, dating back to the first dwellings at Jamestown. Throughout the 19th century, little was done to assure customers of any type of quality in the roofing products they were buying. George Glines and his Patent Slate Paint was a notable exception. Glines would warrant roofs upon which his paint was applied for up to three years, but the paint had to be applied by his men to make the warranty effective.[62]

Little else was done in that respect and the industry suffered accordingly. By the end of the 19th century, quality control in tin and composition roofs was so lacking that the continued existence of each of those manufacturing industries was questionable. Part of the blame lay with the roofing contractors themselves. Some of them were following dubious application practices which could be described as a "how cheap" concept instead of a "how good."[63] This problem even plagued George Washington at his Mount Vernon home. After having some work done there, the roof still leaked, prompting Washington to write: " . . . that there can have been little attention or judgment exercised heretofore in covering it is a fact that cannot admit a doubt; for he must [have been] a miserable artisan or a very great rascal indeed . . . "[64]

The roofing material manufacturers were just as responsible for the condition of the industry at this time as were the contractors themselves, because they were actually advocating bypassing the experience of roofing contractors in their advertising literature by suggesting that people with no experience—the do-it-yourselfers—could apply most roofs.

In 1869, the Mica Roofing Company, in speaking of its canvas-based composition roofs, said that they could be applied by "any ordinary workmen."[65] In 1872, H. W. Johns said that its canvas/felt composition roof was applicable by "any intelligent workman."[66] The 1904 Philip Carey catalog suggested that "honest and intelligent labor is all you require in the application of Carey's roofing. Anyone can put it on who is sufficiently mechanical to handle a knife and hammer." And, "No need to engage a roofer when with Carey's Cement you quickly and permanently repair the damage," Carey said of its roofing cement.[67] The tin roofing industry was not exempt from this indictment either. The Edwards Metal Roofing Company in 1916 claimed its roofing material could be applied by the customer. "You do not have to go to big expense for skilled labor," the company said in its catalog.[68]

The quality and reputation of the roofing industry was in shambles. There were no industry-wide standards and no way of knowing what roofs would or would not perform as promised. The national Association was still too weak, lacking a strong membership base, to do anything more than bring contractors together to talk about the problem.

With the industry in such disarray, the Barrett Manufacturing Company took action in 1906. It employed Alfred W. Erickson to conduct a survey of existing roof systems and to determine their good and bad qualities, why some lasted many years, and why others failed so quickly.[69] Results were compiled and served as the basis for the original "Barrett Specifications" in 1906, a compendium of the good things that the original composition roofers had done. According to the Company, the Barrett Specs "represented the bible of good roofing."[70]

Actually these specifications represented the minimal guidelines to which a standard slag or gravel roof needed to adhere. They called for a minimum weight for unsaturated felt of 5 pounds per square. Single-weight saturated felt was to be no less than 14 pounds per square. Initial nailing was to be done with three barbed roofing nails and at least 120 pounds of coal tar pitch was to be used per square. The gravel was to be no larger than 5/8 inch and was to be no smaller than 1/4 inch. It was to be dust-free and warmed up if cold. If slag was being used, minimum coverage was 300 pounds per square. If gravel was being used, coverage was to be at least 400 pounds per square.

A happy group of union roofers from Chicago celebrate Labor Day, 1905.

The specifications also included the application procedures that were to be followed when installing the new roofs.[71]

The Barrett Manufacturing Company helped revolutionize the built-up roofing industry with these specifications. But mere publication of standards was not enough. In 1916, it introduced the Barrett Specification/Barrett System Roof. The specifications had previously been generic, in that any manufacturers' products could have been used to meet the new guidelines. Now, however, Barrett began manufacturing a complete roofing system that was designed to meet the specifications. To back up their new system, Barrett began offering roofing bonds.

The bond was a revolutionary new concept, but it should not be confused with a guarantee. When a manufacturer issued a bond, he was attesting to the fact that his materials would withstand the elements for a specified number of years. The bonding agency simply guaranteed that the manufacturer had the financial capability to back up his warranty, a fact that only mattered when the manufacturer had gone out of business because that was when the bonding company had to pay off.

Manufacturers also began offering guarantees, but they covered no more than a $5 penal sum per square. This coverage had to be purchased by the customer at $1.50 per square. Because a roof could cost up to $50 a square, one could end up paying a 30 percent premium for only 10 percent coverage.[72]

Notwithstanding these limitations, the Barrett bond for the first time held the manufacturer accountable to some degree for his specifications and material quality. But another factor remained to be considered in the quest for excellence in the built-up roofing industry—the quality of the roofing contractor's workmanship. This was resolved, in many cases, by separate seven-year guarantees that the contractor gave to the customer.

The roofing industry has always been a relatively easy business to enter. A small work force and limited capital could launch a laborer on a career as a contractor. The continual injection of new blood kept the industry vital, but it made quality control difficult. The national Association lacked the strength at the turn of the century to effectively deal with an issue as explosive as certifying qualified roofers. Instead, it was the manufacturers who began to act as industry policemen. This was only natural. Regardless of the specifications and the quality of the material used, a poorly applied roof was doomed to fail. Smart manufacturers did not want their product lines identified with poor performance.

Consequently, roofing component manufacturers were somewhat discriminating about who they let apply their systems. Barrett sought out well-qualified roofing contractors to apply its new system in 1916. They became Barrett Approved Roofers, and were the only contractors that could apply the systems carrying the bond or guarantee.[73]

Johns-Manville realized that "best results are obtained only when the proper roofing is correctly laid." It appointed Approved Roofing Contractors throughout the country, basing these appointments on "thoroughness of workmanship and financial responsibility."[74] In fact, to promulgate a bond, Manville required that a roof be reviewed by one of its inspectors before, during and after application. Richardson/Flintkote took the inspection issue one step further when it authorized its inspectors to "cut the roof as [they] consider necessary," to certify proper application and usage of materials.[75]

The ability of asphalt shingle and built-up roofers to organize both as contractors and as manufacturers was a vital factor in increasing public acceptance of these roofing systems. The willingness on the part of select manufacturers to insist on rigid specifications for applying their materials improved the reputation of those products. However, such tentative steps toward professionalizing roofing had been restricted to prepared roofing manufacturers or slag-and-gravel roofing contractors. The vast range of other roofers who worked with tin, tile, slate or other materials were unorganized and unregulated. The first step toward bringing some order into this branch of the industry did not occur until 1912.

A group of Midwestern slag roofers wanted to broaden the range of roofing materials they applied and thereby increase the range of contracts on which they could bid. To help them select the proper materials and learn the application techniques for systems that were unfamiliar, they formed an association. Initially, there were four members of what they called the Western Roofers Association. However, the organization grew over the next four years to include roofing contractors from New York, New Jersey and other East Coast states. At the next Association meeting in 1916, the membership agreed to broaden the organization to take on a national character. A vigorous new force entered the industry under the title of Associated Roofers of America (ARA). Members were encouraged to purchase only materials that bore the official ARA label. These materials were produced according to rigid specifications laid out by the Associated Roofers of America. To insure that those specifications were fully met, ARA employed inspectors. The organization was a strong champion of the roofing contractor and did not shrink from challenging the quality and cost of the roofing materials' manufacturers. Although ARA was separate from the National Association of Gravel and Slag Roofers, most "progressive" roofing contractors belonged to both groups.[76]

Economically, the last 20 years of the 19th century and the first 20 years of the 20th century were characterized by tremendous economic growth, although there were some setbacks along the way.

Construction expenditures, which mirror these ups and downs of the economy in general, rose from an average of $802 million for the years between 1869 and 1878 to $2.4 billion in 1890. They continued to rise to $2.7 billion in 1892, when the country entered a depression. Expenditures fell over the next four years to $1.9 billion in 1896, but then started climbing again, reaching $2.5 billion in 1900. With minor fluctuations, they continued climbing through 1907 when they totaled more than $4.3 billion. After a decline because of a 1907-1908 depression, they reached $4.5 billion in 1909. With slight digressions in 1910 and 1911, expenditures climbed virtually unhindered until 1920, when they hit the lofty figure of $6.7 billion.[77]

Within this climate of economic growth, roofing contractors thrived. Manufacturers following the example of Andrew Carnegie moved to expand their control over the market. Warren-Ehret was able to create a fully integrated roofing business by orchestrating distinct manufacturing, sales, refining, and application offices. Samuel Barrett emerged as one of the generals of the industry. Already the dominant force in roofing products in the Midwest, Barrett had even more ambitious plans. In 1896, he expanded to the eastern United States. Warren-Ehret was incorporated into the Barrett Manufacturing Company and Michael Ehret became a partner in the larger firm.

Nicholson and Galloway, founded in 1849, still specializes in restoring historal structures. Here, an aerial view of their completed work on St. Patrick's Cathedral in New York City.

Pioneer roofing and chemical companies such as I. D. Fletcher, the Mica Roofing Company, and Warren Chemical & Manufacturing Company were also absorbed. In 1906, Barrett expanded into Canada, absorbing two Canadian manufacturing establishments. When Samuel Barrett died in 1912, his company was the dominant supplier in the roofing industry and one of the most important chemical companies in America.[78]

The need to insure one's profits by changing, expanding and diversifying product lines was not lost on the roofing contractors of the period. The J. D. Candler Roofing Company of Detroit entered business in the 1880s offering galvanized iron cornices as its only product. Sensing the need as well as the ability to create sales, Candler expanded his product line in 1885 to include slate, tin and iron roofing and in 1893 to include gravel roofing.[79]

Nicholson and Galloway in New York, founded in 1849, had concentrated heavily on tin roofing, although the firm did apply slate and tile. As the firm moved into the 20th century, it became apparent that tin roofing was no longer meeting the needs of its customers. The pollution from the coal burned by New York industry was causing the tin roofs to deteriorate rapidly. Consequently, Nicholson and Galloway dropped tin roofing from its product line and substituted composition roofs.

Roofs were pressed into service during World War I for reasons not anticipated during construction.

Perhaps the most intriguing aspect of Nicholson and Galloway's business was the extent to which its two major services diverged. Roofing, of course, was the primary product, but in 1900 the partners expanded into building restoration. Specifically, they restored historic stone buildings, cleaning, performing general maintenance, and waterproofing them.[80]

Economically, this period provided a good opportunity to enter the roofing contracting business. With the exception of 1917, when 24,000 more businesses ceased operation than started, there was an average of 39,000 more small businesses starting up each year than closing between 1900 and 1920.[81] The number of people in the United States who identified themselves as roofers grew from 9,000 in 1900 to 14,000 in 1910. The number dropped to 12,000 in 1920, but that drop is probably attributable to World War I and corresponds to the drop of business entries in 1917.[82] Regardless, the industry was poised and ready to expand further as it entered the 1920s.

During this period of American industrialization and growth, the roofing industry faced and resolved some important issues. The industry acknowledged its role in major urban fires, and worked to provide more effective fireproof roof coverings. With the initiative taken by Barrett, industry members acknowledged the dismal condition of the industry and its reputation at the turn of the century and worked to reverse the trend by embracing the Barrett Specifications of 1906 and the bonded roof. Manufacturers joined in an association to insure higher quality materials. Contractors faced with the dual problems of quality and competition also looked to associations to bring order into the chaotic industry. In forming the National Association of Gravel and Slag Roofers, the Prepared Roofers Association, and the Associated Roofers of America, roofing contractors had taken the first steps toward achieving a stable, modern industry.

Charles Kirberg of St. Louis (now Kirberg Roofing, Inc.) poses proudly with his son in front of his new Model T Ford pickup, circa 1923.

CHAPTER 5

Economic expansion: roofing in the Roaring 20s

"Supine amidst our growing store, we slept securely, and dreamt of more." JOHN DRYDEN

By 1920, the frantic pace of industrialization had transformed America. Internationally, the United States had gone from a debtor nation to the world's banker. The growing American population had concentrated in cities, ending the traditional dominance of rural areas in politics and social life. Even agriculture yielded to the inroads of industrialization. Farm equipment manufacturers produced more than 200,000 tractors in 1920.[1] The rumbling of cars, trucks, and tractors became symbolic of the new progressive decade of the '20s.

John Iran Prilika, a roofing foreman who by 1923 drove a Ford Model T truck to the jobsite, personally experienced the changes of the new era. Ten years before, his employer, the Sherrif-Goslin Company of Battle Creek, Michigan, had been one of the few construction firms in town to own a car and even that company generally relied on a horse and wagon to transport men and supplies to its jobs. But by 1923, trucks had begun to replace the old horse-drawn "Democrat" wagon. This made the jobs of roofers like Prilika much easier. In the past, Prilika always had to be watchful that the hobbled horses did not drop any excreta near the fired tar kettle. The odor from such an accident brought immediate action from the neighbors and even the strong-nosed roofers were repulsed by the smell. A more serious problem occurred if a startled horse upset the tar kettle and spilled the flammable liquid on the burning wood. The sudden burst of flames would bring chaos to the jobsite as men scurried back and forth, some trying to secure equipment from the flames, others running for the fire department. This type of accident, which could threaten the building being roofed, seldom won the contractor his customer's goodwill. A seemingly small change in the method of transportation made the contractor's job easier.[2]

Unfortunately, not all the changes that had occurred were so positive. The roofing industry's leadership had undergone a transition during the 1920s; the old guard had passed on. The industry had expanded rapidly in the second half of the 19th century, particularly in the area of composition roofing. But the young men who had led that expansion had passed from the scene by 1920. Samuel E. Barrett died while returning from a business trip in December 1912. Michael Ehret, who had invented slag roofing, died a year later at the age of 79. Moses W. Powell, founder of the first trade association, died in 1907.

By 1920, many family-owned roofing companies had faced the crisis of leadership change. Where no children were present to carry on the businesses, roofing firms often became defunct. For more fortunate firms, the transition was smooth; one generation gradually replaced the next.

The Wehner Roofing & Tinning Company of Dayton, Ohio passed from its founder, August Wehner, to Wehner's son Joseph in 1885. By 1920, another uneventful transfer had taken place and Vincent Wehner took the helm of the family business.[3]

For M. W. Powell and Company, the transition was not that graceful. The immediate successor to Moses Powell was his son-in-law, Victor Barbour. He continued his predecessor's involvement in the national Association and initially, provided the firm with steady, if uninspired, leadership. Unfortunately for Barbour, Moses Powell had a grandson by his second daughter. The young man, Myron W. Powell, was a poor administrator of both personal and corporate funds. He exhausted his considerable personal capital in short order and began dipping into the company coffers. By the end of 1922, he had lost even those monies borrowed from the company, and began hocking the roofing firm's stock to pay debts he had incurred.

The Board of Directors, led by Barbour, engineered a bailout for the young man. The company had been gaining "unpleasant notoriety through the daily press." Because Powell's initials were the same as the corporate name, clients and creditors alike had begun to question the solvency of the roofing firm. The young man was twice relieved of his debts by the company. The final occasion saw him removed from the Board and all his certificates of corporate stock seized as security for repayment.[4] This humiliation seems to have shaken him, and he gradually began to stabilize his lifestyle and finances. But it was not until 1926 that he could pay his debt to the company in full.

The J. D. Candler Roofing Company of Detroit lost its founder in 1925. As a young man, James Candler had declined the opportunity to join his father's business and developed his own firm into one of Detroit's leading metal roofing businesses. When he died, his sons were willing to join the business, but it was his wife, Maria Candler, who took up the reins and became company president. She personally guided the firm through the remainder of the 1920s.[5]

The John Sykes Company, which was founded in 1901 in Atlantic City, also saw its founder retire in the 1920s. Three sons jointly shared the management responsibilities for the firm. Although their father's business had originally been very small and was operated out of an attic office, the brothers expanded the company greatly. Specializing in slate, tile, and tin roofs, they benefited from the development of Atlantic City into an exclusive resort center.[6]

A booming economy was smoothing the transition for older roofing companies. During the 1920s, the construction industry and the automobile industry led the entire nation through a decade of unprecedented prosperity. There were sectors that lagged behind, notably textiles and, to a lesser extent, coal mining. But overall, it was a time of plenty; nothing suits the building industry as well as economic expansion.

Construction expenditures had increased steadily between 1912 and 1920, despite the intervening war years. But because the government drove up interest rates by borrowing to finance the war, and because of the diversion of labor and material to war industries, construction expenditures had not risen as fast from 1917 to 1920 as they otherwise would have. From a 1920 high of $6.7 billion, expenditures fell back slightly in 1921. This reflected a small post-war recession. With a recovery in 1922, however, expenditures doubled the 1920 rate; by 1926 they had reached $12 billion. This dramatic growth represents the market's attempt to catch up with the pent-up demand created by the construction restrictions of the war years. Construction expenditures basically leveled off at that point, actually falling slightly to more than $11 billion in 1929.[7]

This period of tremendous construction growth provided a good opportunity for people to enter the contracting industry. The amount of capital available to a potential business owner was frequently a limiting factor in determining what business he would enter and how much of an investment he would make. It follows that businesses that did not have high start-up costs were the easiest to enter. In this respect, construction work in general and roofing contracting in particular were ideal fields of endeavor. A truck, kettle, mops and assorted tools were all that were necessary to start a roofing contracting operation. If an owner was going to start working too, perhaps no more than two or three additional men were needed to complete a small crew. The balance of the material needed for a job—the asphalt, felt and gravel or slag—would be acquired on a job-by-job basis. There was no need to tie up large amounts of capital in that type of stock.

Fields that required little capital and carried low overhead were the easiest to abandon when times turned bad; with little invested, there was less incentive to persevere. In 1929, for example, construction companies valued their equipment at 7 percent of the year's total business. Overhead was only 14 percent of the year's gross. Thus, as it was succinctly stated, the builder " . . . is under no particular compulsion to continue operation when rents and prices decline appreciably below average total unit cost. He simply drops out of business."[8]

The great majority of roofing concerns entering the industry were small businesses. It is appropriate, then, to examine the entry/exit figures of small business in general and, by extrapolation, apply them to the contracting and roofing industry. This method seems proper, as the figures virtually mirror construction industry expenditures. In 1921, 56,000 more firms entered business than exited. This figure no doubt represents a number of war veterans who could not find work upon their return from Europe. In 1922, the differential dropped to 13,000 more entering than leaving; the decline was most likely a result of the 1921 downturn. However, from 1923 through 1925, there was a minimum of 45,000 more firms per year starting up than closing down. The differential leveled off at that point; it was 13,000, 27,000 and 13,000 for the years 1926 to 1928.[9] The number of people identifying themselves as roofers also doubled, from 12,000 in 1920 to 24,000 in 1930,[10] reflecting the doubling in construction expenditures from 1920 to 1930, as well as the overwhelming differential favoring those entering business as opposed to leaving it.

The decade saw new blood enter the roofing business. Immigrants and the sons of immigrants began to join the ranks of the contractors. Konstantine Karis was a typical example. He established a roofing and metal business in the Richmond section of northeast Philadelphia. The area was the industrial workshop of the city. Small machine shops abutted great chemical, paper and steel plants. Ethnic workers from throughout Europe gathered in small neighborhood enclaves. It was an excellent environment for Karis to make his business a success. He had little in the way of capital when he began in 1920, but he was able to make the most of his small investment by building his own kettle. He kept his business profitable by performing metal work for some of the small shops in the neighborhood and even dabbled in

A tornado rocked Indianapolis in May 1927, but a shingle manufacturer found a silver lining. The caption on this photo: "Even a cyclone cannot unlock the Ford Cyclone Safety Shingle."

heating and insulation contracting. Gradually the prosperity of the times and Karis' persistence placed his roofing business on a profitable basis.[11]

Throughout the country, roofing contractors were expanding their operations. Former workers were trying their hand at contracting. For many new roofing contractors, the strength of family ties held their businesses together in the first few difficult years of operation. When J. Roy Martin of Anderson, South Carolina, began his roofing business in 1928, he had little more than debts and optimism. He borrowed $2,000 on a life insurance policy, which he invested in a Model T truck and roofing tools. But Martin did have the help of his two brothers, who became the core of his work force. As contracts began to materialize, he was able to hire a part-time laborer, but he also began to use his 10-year-old son when the boy was off from school. With this level of family involvement, the business was able to grow from a basement storeroom gamble to an established contract construction firm.[12]

Figures from some of the major roofing contractors demonstrate that the 1920s was a profitable era. On October 23, 1922, A. E. Chevalier and a group of investors purchased Warren-Ehret's independence from Barrett for $138,835.[13] After six years on its own, and after the post-war building boom had peaked, the firm achieved gross sales of more than $1 million, of which net profit was $63,512 or 6.3 percent of gross sales.[14]

Judging from dividends paid, M. W. Powell also had a very lucrative period during the 1920s. In 1926, at the conclusion of the construction boom, Powell paid a 100 percent dividend. Dividends were limited to 15 percent in 1928 and 1929.[15]

The J. D. Candler Roofing Company entered the '20s in a unique posture. Perhaps because of the increase in roofing competition between 1920 and 1930, Candler cut salaries by 20 percent to help reduce overhead.[16] His success is presumed, as Candler turned a $6,230 profit on sales of $143,441 in 1924. Profitability continued and in 1927, a $2 per share dividend, the first since incorporation, was declared. Perhaps most telling of Candler's success was a 1929 $6,124 bonus voted to employees.[17]

A feature of the general strength of the roofing industry during the 1920s was the emergence of a single industry-wide association. The National Association of Gravel and Slag Roofers, founded in 1887, and the Associated Roofers of America, founded in 1916, both served much of the same constituency. The leaders of the top organizations had long wanted to unify their efforts. In 1921, they arranged for both organizations to hold their national conventions in Chicago simultaneously. Sentiment for a merger grew during the three days of working sessions and festive receptions. On January 19, 1921, the directors of the two associations met and voted to form the United Roofing Contractors Association of North America (URCA). The title of the organization reflects the involvement in the merger of Canadian contractors. For the first time, an association represented the combined interests of tile, metal, and composition roofing contractors. Siding and waterproofing contractors were also involved. The unified industry entered the 1920s with confidence.[18]

One of the principal forces behind the formation of the United Roofing Contractors Association was Eugene M. Pope. He was founder of the principal trade publication of the industry, *American Roofer*. Initiated in 1911, the publication had long worked to improve the industry and protect it from "the enemy within"—the inefficient and inexperienced contractor. Articles on proper business methods and the latest application techniques, or explanations of the intricacies of Workmen's Compensation, served to increase the efficiency of the roofing contractor. Pope's integrity and independence made him an important middleman in the negotiation that led to the unification. He also became one of the most important officers in the United Roofing Contractors Association. Pope served

Eugene M. Pope, founder of American Roofer *magazine and its editor from 1911 to 1934.*

as the secretary-treasurer of the Association until a few years before his death in 1934. He provided URCA with the type of stable leadership it needed in the fast-paced economy of the 1920s.[19]

A general rise in wages contributed to the overall prosperity of the period. The average worker's pay doubled between 1914 and 1923. This was partially because of the war with Germany. The effect of such a jump was to increase the average American's purchasing power and expand the demand for consumer goods. The higher wages also undercut the worker's best reason for joining a union. After three decades of growth, the American labor movement began to lose ground in the 1920s.[20] But that is not to say the period was a peaceful period in labor/management relations in the roofing industry.

Initially, workers in the roofing industry were divided by both geography and the type of roofing systems they applied. Slate and tile roofers belonged to different unions than composition or metal roofers. The danger of these divisions to the workers was demonstrated by a 1906 job action in New York City. That year was a busy one for contractors as several multi-union complexes were under construction. As these buildings neared

The F. J. A. Christiansen Roofing Company, Inc. in Milwaukee was quick to note the advantages of the truck over the horse-drawn wagon. Christiansen maintains this 1916 Model T today as part of its company collection.

completion, the local tin roofers' union struck for higher wages. The builders were frantic to complete the structures and open them for use but the strike dragged on for months. Finally, the general contractors decided to put composition roofs on the buildings as a temporary measure. Composition workers crossed the picket line and applied the roofs. Coincidentally, the New York Board of Fire Underwriters chose this time to withdraw its 100 percent differential for fire insurance rates for buildings covered with composition systems. This further increased the popularity of composition roofing. The tin workers won their strike after a long battle. But the lack of solidarity among composition and tin workers had led to considerable reduction in the volume of work available to the returning strikers.[21]

The only true worker solidarity that cut across the various branches of the industry in those early years was shown by the Roofing Teamsters' Union. But the growth of the American Federation of Labor, which by 1920 had 3.2 million members coast to coast, provided a mechanism for bringing workers together.[22] Building trades workers were the core of the American Federation of Labor's strength. Unlike other industries, building trades, like roofing, had been protected from many of the swift changes industrialization brought to factories and workers. In spite of jurisdictional battles among the different trades, the workers in many localities were able to work out collective bargaining procedures.[23]

Under the guidance of the American Federation of Labor, the United Slate, Tile and Composition Roofers/Damp and Waterproof Workers Association was established in 1902. After an initial surge, membership leveled off at about 1,800 and was maintained at that level through 1920. With the upswing in construction expenditures, the union's membership leaped in 1921 and 1922, reaching 3,000. This represented 25 percent of the 12,000 people who identified themselves as roofers in the 1920 census. Union membership remained constant until at least 1925, while the number of people identifying themselves as roofers steadily increased.[24]

In some ways, the roofing union was among the least successful in the building trades. In absolute numbers, the roofers' union lagged behind carpenters and painters, but their union also lagged behind plasterers, sheet metal workers and cement finishers. More to the point, roofers were among the lowest-paid building trades workers in the 1920s. In a 1923 survey of Phildelphia building tradesmen, the average pay for a roofer was revealed to be 65 cents per hour. Electrical workers, sheet metal workers, and steamfitters were paid a full 25 cents per hour more.[25] No doubt the reason for the low salaries and small union membership was that roofing required a lower degree of skill than carpentry or masonry work. Because of the low skill levels involved, many workers made less of a commitment to the trade than they would have if it demanded a long apprenticeship. Where workers did support the union, they received significantly higher wages than non-union workers. In New York, for example, members of the United Slate, Tile and Composition Roofers Union received $1.25 per hour, while non-union workers averaged $1. In New Orleans, 1924 statistics show that union workers were paid more than 30 percent more per hour than their unorganized counterparts.[26]

Although the 1920s have been described as "the lean years" of the modern labor movement, the United Slate, Tile and Composition Roofers Union adopted an aggressive posture during the period. Contractors, on the other hand, followed the lead of the National Association of Manufacturers and pushed for an open-shop roofing industry. Lines were drawn and a battle was fought on a nationwide front.

J. T. Hurley, the union's principal organizer, was leading the fight. He was an energetic man who spent much of his time on the road, spreading the gospel of trade unionism. His job was to organize new locals, revitalize dormant ones, and adjudicate jurisdictional disputes. Because of Hurley's fieldwork, the United Slate, Tile and Composition Roofers Union expanded from 75 locals in 1926 to almost 100 in 1929.[27]

Hurley met with failure in some of his efforts. In some strongly union cities, such as Detroit, he was met with apathy on the part of the roofing workers. In July 1926, he tried to get all the non-union roofers to attend an open meeting to explain what the union could do for them, but only a dispirited few attended.[28] In May 1927, he called an organization meeting in Fort Worth. Initially, there seemed to be a good response from roofing workers and contractors alike, but when a second meeting was held, many supporters who had earlier offered help failed to come through.[29] In Long Beach, California, the contractors seemed to have no problems with their men forming a union. The problem was the workers themselves. Most agreed that a union could help their conditions, but they resisted Hurley's attempt to do the initial organizing.[30]

Milwaukee presented another problem. Hurley arrived there in June 1927; he was to try to reactivate a long-dormant local. He found that all the old officers had left town and that most of the large contractors were members of the Open Shop Association; he did not have much success on that trip. Perhaps the most significant factor that worked against organizing a new Milwaukee local was the distrust felt by the rank and file. The treasurer of the previous local had absconded with much of the union's funds.[31]

When Hurley visited Syracuse in 1928, he found 23 roofers interested in organizing a local. Their plans cooled significantly when it became known that virtually all of them were paid on a steady-time basis. Much to Hurley's dismay, they were afraid they would jeopardize this arrangement if their employers found out they were forming a union local.[32]

Hurley met with resistance from contractors across the country. Their attitude was summed up by a Hartford contractor, who declared emphatically that he would not hire union men "if (I) can get along without them."[33]

As the organizer scurried back and forth across the country, he also acted as a mediator, trying to solve disputes between his locals and the contractors in various cities. His mediation attempts were also met with mixed success. In Youngstown, Ohio, the contractors agreed to a contract that established an eight-hour day. It paid time and a half for all daily overtime up to 9 p.m. and double time for work thereafter, as well as on weekends and five legal holidays. Roofers' wages were to hold at $1.30 per hour for a fixed time, and were then renegotiable.[34] In Cleveland, a contract was negotiated for 20 separate trades. Workers agreed to keep wages at $1.30 for 1926, but were then to get a raise to be within 25 cents of an average of 20 other trades listed in the contract.[35]

Other mediation efforts met with less success, particularly when Hurley faced united contractor opposition. In 1926, Local 79 in St. Louis went on strike for a 25-cents-an-hour raise. The contractors were adamant; they would not even discuss it. The union then reduced its demands to 17.5 cents an hour. The contractors listened but still would not accept it. With the help of the organizer, they finally accepted a raise of 10 cents an hour with the contractors agreement not to discriminate against their local members.[36]

Hurley's most difficult challenge came in Florida. The southern half of that state, particularly cities such as Tampa, St. Petersburg, Coral Gables, and especially Miami, rode the crest of national prosperity. Throughout the early and mid '20s, real estate prices soared in the Sunshine State. Construction boomed as hotels, private homes and guest cottages sprouted out of the marshes. The boom in south Florida was based on inflated real estate values and unrealistic expectations of future growth. Tourists did flock in at an astounding rate. But when the winter of 1925-26 was unseasonably cold and rainy, the image of the Sunshine State clouded. In a matter of months, the great Florida land boom was over and the state's economy was in shambles. Loans were called in, banks failed, paper fortunes disappeared, and solid businesses were threatened. The Depression that would cripple the nation in 1929 began three years earlier in Florida.[37] A tainted environment for industrial relations followed in the wake of the collapse.

Economic hard times were the greatest threat to union strength. Men desperate for jobs were always willing to undercut striking workers. Contractors trying to survive the crisis were interested in slashing labor costs, not inflating them. In May 1926, the Miami local of the United Slate, Tile, and Composition Roofers Union struck for a wage increase. Contractors were outraged and acted decisively. A Master Roofer's Association was formed to create a common front to the union's demands. After a two-week strike, Hurley was able to pull off a minor miracle. He convinced the McDonald Lumber and Roofing Company, Florida's largest roofing contractor, to agree to a one-year trial contract with the union. He negotiated a two-year wage plan with the Master Roofer's Association that would raise wages by 25 cents per hour between 1926 and 1928.[38] The roofers went back to work on August 27, 1926.[39]

The union did not restrict its action to conflicts with management. One of its most vexing issues was its attempt to ward off the encroachment of other trades. There were continuous battles with carpenters and their union about who would apply asbestos shingles. Sheet metal workers were occasionally found applying slate, tile and composition roofs, although they were authorized to install only metal roofs. There was a bitter dispute in 1928 with the Brotherhood of Painters, Decorators and Paper Hangers when it was discovered that its members were spraying waterproof coatings on some jobs in Sacramento, even though that was work that was normally reserved for the waterproof workers of the union.[40]

Not all dealings between the roofers' union and contractors' associations were confrontational. Cooperation could be effected when mutual interests were at stake. The classic example of this was the 1920s fight over roofing insulation installation. In 1925, an appeal was made by the carpenters to the National Board of Jurisdictional Awards to take the installation of fiberboard installation away from the roofing trades. Within a matter of weeks, the United Roofing Contractors Association under its president, George Moore, assembled a delegation of contractors from around the country. They met with George Jones, president of the roofers' union, and formed a unified defense of the roofing trade's right to install rigid roofing insulation. This two-fisted approach warded off the attempted encroachment and laid the foundation for insulation to become an important part of a contractor's trade in the years to come.[41]

Although roofing insulation came into its own during the 1920s, it evolved from many years of experimentation. It was the problem of condensation that first connected the roofing industry to insulation. Moisture collected on the underside of many roofing systems, especially tin or composition roofs. Some contractors tried to avoid this problem by placing a layer of felt paper under the tin roof. In 1910, Joseph Benn, a Rhode Island contractor, had a better idea. He laid cork insulation over the roof of a dye house in the town of Greystone. This experiment pioneered the way for the use of cork board as a roofing insulation, but it did not end the evolution of insulation systems. Some Midwestern industrialists experimented with the use of sawdust to make fiberboard insulation, while others turned to flax. Both attempts met with some success,but it was bagasse, the waste of sugar cane production, that proved to be the breakthrough for roofing insulation.[42]

The man behind this innovation was Swedish immigrant Bror Dahlberg. Dahlberg left his native land at the age of 10 for the United States. Like many a Swede before him, he headed for Minnesota. He worked a series of odd jobs, beginning at a wage of $2 per day. By the time he was 27, he was manager of Harwood Furniture Factory. He later became involved in paper manufacturing. This brought Dahlberg into contact with the lumbermen who were experimenting with fiberboard insulation. The young Swede's advantage in the race to improve insulating materials proved to be his contacts in the Louisiana sugar cane industry. As was the case with so many roofing industry innovations, Dahlberg was able to take a waste product that could be had for next to nothing and transform it into a valuable innovation. In this case, the waste product was bagasse—cane fibers crushed to extract juice for sugar.[43] Dahlberg used the bagasse to make a high-performing, lightweight roofing insulation he called Celotex. To control his source of raw material, he founded the Southern Sugar Company in Florida, which became the basis for the sugar industry in that state. Meanwhile, in Chicago, he built his Celotex plants and supervised the marketing of the synthetic board. His success marked the

Bror Dahlberg, founder of Celotex Corporation. Dahlberg found a use for cane fibers that had been crushed to extract the juice for sugar.

beginning of a multi-million-dollar company and the solidification of insulation as a branch of the roofing industry.[44]

Although the 1920s was a period of conflict between old and new, there were no major technological innovations in the application of composition roofing. Contractors relied on proven methods and materials. Roofing contractors still brought the pitch to the jobsite in wooden barrels. Once the tar had been put in the kettle, the barrel staves were burned to heat the pitch.

The equipment at the jobsite was a jerry-rigged tribute to ingenuity and frugality. A hay wheel would be fastened to a crudely made derrick, which was anchored by several burlap bags of gravel. The heated pitch traveled up this device in 5- or 6-gallon buckets. Rolls of saturated tar felt were then spread on the deck, using homemade mops of cotton yarn. In the 1920s, composition roofing contractors continued to keep alive the old tradition of employing homespun equipment and supplies.[45]

Although built-up roofing was technologically static during the 1920s, the period was one of the most innovative for shingle roofing. Asphalt shingles were first manufactured by Henry M.

A 1920s Barrett advertisement boasts the surety bond behind each Barrett roof.

Reynolds, a Grand Rapids, Michigan roofing contractor and manufacturer. In 1903, he made shingles out of stone-surfaced sheets of roll roofing. The shingles were cut by hand with a simple knife. Although the size of the shingles varied, because the worker doing the cutting often became bored and careless, they were generally 8 by 16 inches.[46] Incredibly, Reynolds did not patent his invention, but invited other manufacturers to copy his idea. Most did not. They were interested in perfecting the use of roll roofing. When the long rolls of saturated felt were cut into shingles, they usually became susceptible to wind damage. But in 1914, an innovative roofing mechanic named F. C. Overbury of the Flintkote Company experimented with applying crushed granules of slate to the asphalt shingle. This helped hold the shingles in place when they were exposed to rough weather. Overbury also experimented with cutting both roll roofing and shingles into various patterns that would enhance the roofs' appearance and durability.

World War I stimulated the use of asphalt for residential roofing. Its popularity increased because it used non-strategic materials and was easier to ship than wood or slate shingles. But it was not until the 1920s that asphalt shingles became a major part of the roofing industry. The period saw significant improvements in the aesthetic effect of a shingled roof. The variety of colors available increased and so did the range of shapes. Shingles shaped in hexagonal, rectangular, octagonal, even circular styles were employed, sometimes all on the same roof. Residential builders began to see the advantages of the textured effects of a shingle roof. They also appreciated its low cost and fire resistance. By 1929, the asphalt shingle had become firmly established as "Everyman's Roof."[47]

The spread of asphalt shingles was not without its troubles. Throughout the 1920s, some unscrupulous manufacturers insisted on dumping asphalt shingles on the market that failed to meet ARA standards. These seconds looked perfectly all right to novices and were sold to contractors willing to palm them off to unsuspecting consumers. Another threat came from the mail-order business. Both roll roofing and asphalt shingles appeared to be simple systems to apply. Several mail-order businesses took advantage of this and sold asphalt shingles directly to the consumers—assuring them that by applying the material themselves they could be assured of cheap, waterproof roofs. Of course, what often happened was that the improperly laid shingles failed to protect the house and the disgruntled homeowner loudly blamed the shingles. The United Roofing Contractors Association and the *American Roofer* vigorously defended the industry's interests. Through editorials and advertising, they promoted awareness among contractors and in the building industry of the danger of these profitable but unprofessional practices.[48]

Technical changes in the industry always seemed to be accompanied by quick-buck hucksters. The roofing industry began to see the development of a wide variety of new roof coatings in 1926. Some of these were the result of the increased sophistication of the manufacturing sector in the post-war era. Industrial laboratories became partners with all large manufacturers, especially in fields like roofing materials that had always been related to the chemical industry. The trouble was that many new products in the waterproofing field were bringing men to the business who were unskilled roofers. Ambitious laborers, carpenters, and apprentices were seen by roofing contractors not only as a competitive threat, but as a threat to roofing's already tenuous reputation. In the face of this challenge, established roofing contractors in many cities across the country began to form local associations. Detroit, New York, Portland, Salt Lake City and numerous other cities and states saw new local organizations emerge.[49]

The United Roofing Contractors Association, unable to deal with the issue itself, encouraged the formation of such local associations. URCA could identify and publicize problems, but it lacked the executive power to do much more. What the Association did best was to educate the roofing contractor. It was through URCA that the most important business marketing tool of the 1920s, installment selling, spread quickly throughout the industry. Under this plan, the homeowner in need of a new roof could spread the cost over several months, thereby making the burden of reroofing less oppressive. It also encouraged the homeowner to use a professional roofing contractor rather than be seduced by the mail-order advertisements that seemed to offer a cheaper route to home repair.[50]

The 1920s was a period of prosperity for the roofing industry. There were more roofing contractors, more roofing workers, more local industry associations than ever before. Old roofing systems prospered; new ones were perfected. Yet in these good times, nagging problems arose. Labor relations were uncertain, quality control was not perfected even among concerned contractors, and roofing still suffered a decidedly second-class status in the construction industry. Such problems could be ignored when the profits continued to accrue, but by 1929, roofers began to feel the chill of a business downturn. Hard times and major challenges lay ahead.

Attendees at the 1938 URCA convention in St. Louis look calm despite the fact that the Association had, organizationally speaking, hit rock bottom. One executive secretary was fired and another hired during the meeting.

Coping with hard times: the Depression's effect on roofing

"Everything in the world may be endured except continual prosperity." GOETHE

"Wall Street Lays an Egg," read the headlines of *Variety* on October 29, 1929.[1] Overnight, the nation's financial system collapsed. More than $30 million was lost when more than 16 million shares of stock were sold in a panic. The companies that had greedily directed their profits away from reinvestment and to the market's margin funds lost badly needed capital. Banks that had made loans to speculative investors suffered tremendous losses.

Banks failed, taking with them the life savings of many hard-working, thrifty Americans, some of whom probably didn't know what the stock market was, much less speculated on it. Unemployment was widespread. More than 100,000 people were losing their jobs every week in 1931, and by 1932, national unemployment had victimized 25 percent of the nation's population. It was worse in some areas. In steel towns, unemployment reached 80 percent of the population. Coal mining towns offered virtually no employment. By 1932, the originator of the 1928 Republican campaign slogan, "A Chicken in Every Pot and Two Cars in Every Garage," was himself unemployed and a beggar.[2]

The stock market crash caught the nation totally by surprise, but it shouldn't have. The indications of it were present. The Florida land boom climax had demonstrated as early as 1926 that speculative cycles could not be sustained indefinitely. Throughout the late 1920s, the nation's wealth was concentrated in fewer and fewer hands. consumer spending, so prominent in the first part of the decade, began to decline. Automobile sales dropped; inventories began to accumulate, and in the summer of 1929, construction volume began to fall off rapidly. All of these were signs that problems lay ahead, but Americans paid little attention. Banks kept lending and market speculation continued. On October 24, after several months of wavering, the market fell dramatically. But with the cooperation of the J. P. Morgan Company and other major financiers, enough money was pumped into the system to bolster it, at least for a few more days.

For most roofing contractors, the Depression was not a shock. There had been signs of trouble for nearly a year before the crash. Unfortunately, as the pace of construction slowed, contractors hoped to maintain profits by increasing their volume. Prices were cut to increase sales in the face of stiff competition. The resulting price-cutting spiral was inexorable. Profits made in earlier years were swallowed up. Manufacturers and contractors both felt the effects. In 1928, many companies, such as CertainTeed Corporation, a major asphalt shingle producer, had to declare losses for the first time since its inception in 1904. The United Roofing Contractors Association was helpless to do anything but condemn price-cutting wars.[3]

The Depression was a merciless experience. Those who had lost their jobs began to lose their houses. Farmers, a group that had not shared the prosperity of the '20s, saw their meager returns dwindle to nothing. They, too, began to lose their properties to banks and creditors, thus prompting a 1933 pledge from Iowa farmers to lynch anyone attempting to foreclose on one of their own.[4] Many were looking for work, but the available jobs did not come close to meeting the demand. Birmingham, Alabama offered 800 laborer positions that paid $2 a day for 11-hour days; the city received 12,000 applications. In New York City, about 5,000 people applied weekly for 500 menial labor positions.[5] Many local governments, civic organizations and churches took on the Herculean task of trying to feed and shelter the victimized. Soup kitchens with lines three and four people abreast and several blocks long were the manifestation of that effort. People lined the streets selling pencils, apples, anything that could bring a few pennies. "Buddy, can you spare a dime" was not only a song that had swept the country by 1932; it was a refrain heard frequently on the streets of America.[6]

The Depression gave birth to several new terms. "Hoover flags" were empty pockets turned inside out. "Hoovervilles" were the towns built of scrap lumber and cardboard on the outskirts of major cities where the homeless and unemployed were forced to live. "Hoover Pullmans" were railroad box cars, the form of transportation used by many to search the country for jobs. And a "Hoover blanket" was an old newspaper, used to shelter its owner from the weather as he or she slept on a park bench.

Describing the situation in Los Angeles, a member of Local No. 36 of the roofer's union related that " . . . business has gone from bad to worse and since June 1st, the bottom has completely dropped out; the Hoover flag is more in evidence now than at anytime during this Depression and as a man said to me today on a job, he says I don't see how we keep living the way we work — about one day every two or three weeks . . . "[7] He went on to relate that union men traveling around the country said, "We're better off here than any place they visited . . . so I guess we won't kick although it is getting worse every day . . . so don't let anyone kid you as the worse is yet to come if you haven't already had it."[8]

The Hession Roofing Company truck is a rolling advertisement for the craftsmanship of its owner.

Throughout 1930 and 1931, J. T. Hurley, the union organizer, continued to travel around the country. He was still trying to organize new locals, but more of his time seemed to be spent resolving problems, especially jurisdictional ones. In Washington, D.C., for instance, he found bricklayers on a government job installing promenade tiles on a roof. The bricklayers claimed it was because they didn't think the area had any union composition roofers and they did not want the work done by non-union workers. Hurley pointed out that there were three union locals and that promenade tiles set in mastic, as these were, were the roofers' responsibility. The bricklayers wanted to divide the work but the roofers remained adamant.[9] In other disputes, he confronted carpenters and sheet metal workers, both groups applying types of roofs that building and trade councils had awarded to the roofers. None of these disputes were resolved quickly, each party reluctant to give up any work. Each trade union was concerned more about the ability of their respective members to work than union solidarity.

The more construction expenditures declined, the more jurisdictional disputes seemed to pick up. From a 1926 high of $12.5 billion, expenditures slid to $11.2 billion in 1929. They then fell to $9 billion by the end of 1930, and plummeted to $3 billion in 1933.[10] A corollary was the number of firms going out of business as opposed to those just starting. In 1931, 49,000 more firms failed than started; in 1932, the figure was 116,000 more.[11]

The initial reaction of the roofing industry's principal trade publication, *American Roofer*, was calm. Editor Eugene Pope had seen the Panic of 1873 and the business depression of the 1890s and was not going to be stampeded into alarmist rhetoric. As late as January 1931, he emphasized the positive effects of the Depression:

"One important result of the lack of business during 1930 in the roofing trade, as in other lines, has been the winnowing out of weaker concerns to the benefit of the stronger. We have just closed a year in which the shaking down and out process in the roofing business has been especially noticeable."[12]

But as the downturn continued, even well-established firms began to buckle under the strain. CertainTeed Corporation was forced to retrench. Plants were closed, wages cut, and prices slashed in an effort to increase profitability.[13]

M. W. Powell and Company seems to typify the plight of the roofing contractor. The year 1930 was characterized by "very poor business and very keen competition." Yet a very small profit

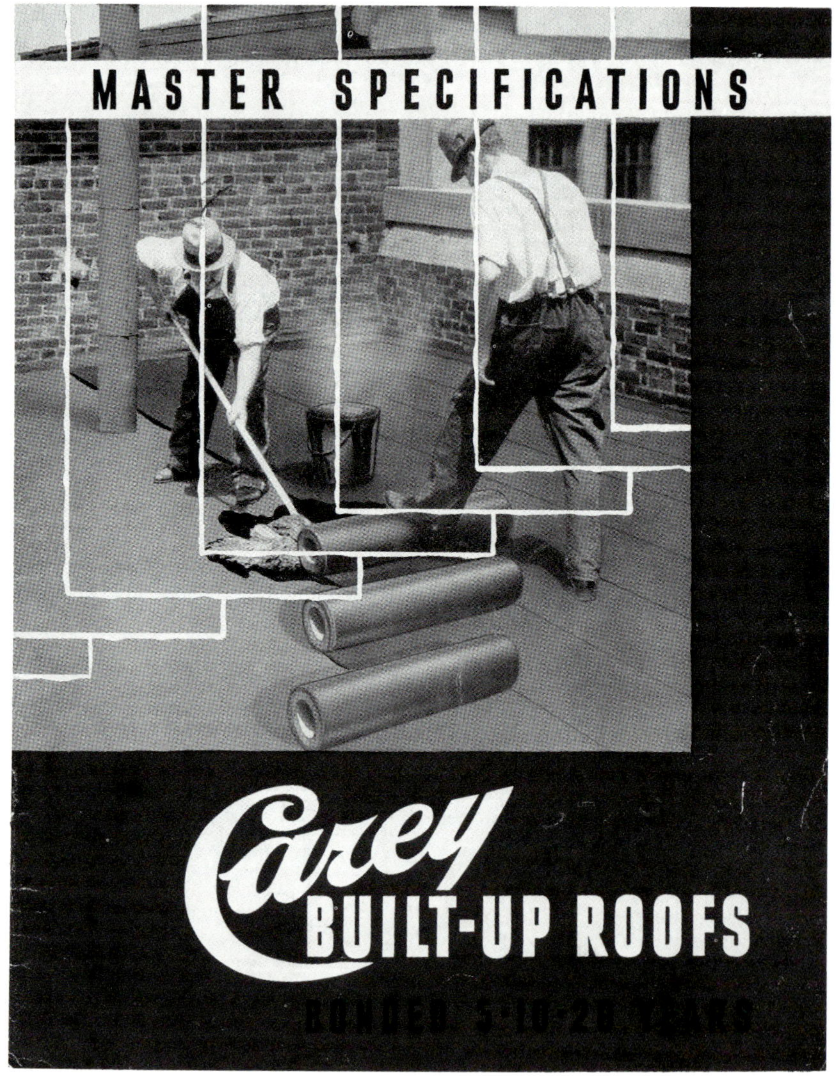

The Philip Carey Company's 1938 edition of master specifications for built-up roofs.

was made and an 8 percent dividend was paid out of the profit and a surplus account. In 1931, "owing to the small volume of business that was done in 1931, with attendant low prices caused by the very keen competition that prevailed during the year, the company, for the first time in a long period of years, showed a net loss as a result of its year's work." A significant indication of the situation's severity was the fact that no dividend was paid in 1931. There was an $8,000 loss in 1932 because of the "natural result of the existing general Depression, with attendant keen competition and consequent low prices."[14] Powell continued to operate at a loss through 1935.

Warren-Ehret paid a 3.5 percent dividend at the end of 1930 and in July 1931.[15] The company acknowledged the severity of the situation and reduced costs, but, believing the Depression was not going to last, it continued to pay dividends.

With construction expenditures dropping quickly, Warren-Ehret reduced salaries 10 percent on July 1, 1931; they had been reduced a total of 33.3 percent by March 16, 1933.[16] The average manufacturing worker's salary was reduced about 32 percent over the same period, from an average of $25 a week to $17 a week.[17]

The competitive struggle for the decreasing roofing dollar made estimating an increasingly important part of the contractor's work. Earl Morrison, who had been a Candler employee since 1899, found this to be the major difference in how the Depression affected roofing. He later recalled:

"In the early days, there were many customers whose work we did every year, when we could get around to it. Nowadays [late 1930s], it is a matter of figure and refigure jobs until you reach the customer's price and keep hustling to get enough work ahead to keep the construction crew happy."[18]

To survive cash flow problems and jobs that were bid too low, the Candler Board of Directors established a $10,000 line of credit with a Detroit bank.[19]

Candler's cost-reduction efforts were not as drastic as some, but in 1932, the Board directed Vice President W. Wallace Candler to devote the "majority of his time to actual estimating and securing of new contracts."[20]

Lawson Roofing in San Francisco is another example of a firm that cut back its expenses to a bare minimum. The operation was reduced to one truck and crew operating out of an apartment building basement.[21]

The Depression caught many manufacturers off balance. The era was marked by a series of corporate take-overs. The Ruberoid Company, a roofing products firm organized in 1886, purchased several rivals during the price-cutting wars of 1928-1929. The H. F. Watson Mills, which had manufactured built-up roofing materials since before the Civil War, and the Continental Roofing Mills of Massachusetts and Maryland came under Ruberoid's control.[22] In 1938, CertainTeed Corporation, the biggest manufacturer of asphalt roofing materials, fell to the management of Celotex, the insulation giant.[23] The high capital demands of asbestos production made firms involved in this branch of roofing and insulation materials particularly vulnerable. In 1930, Ruberoid bought a controlling interest in one of the asbestos pioneers—Eternit, Inc. Again in 1937, the aggressive Ruberoid managers took the offensive. They acquired control of the Vermont Asbestos Corporation, then the only asbestos mining property in the United States. Direction of Keasby and Mattison, which introduced the first asbestos shingles, was taken over by the British partnership of Turner and Newall in 1934. Within two years, Keasby and Mattison became a wholly owned subsidiary of the British firm.[24] Although there were manufacturers who were forced out of business, most reduced costs, redoubled sales efforts and hoped for better days.

"Let me assert my firm belief that the only thing we have to fear is fear itself."[25] Spoken at his inauguration on March 4, 1933, newly elected President Franklin D. Roosevelt set about the business of pulling the country out of the Depression. Within the first hundred days of his administration, he called a bank holiday to allow the entire banking system a chance to stabilize. He also proposed, and pushed through Congress, a number of measures to put America back to work. Referred to as "alphabet soup" because of the way they were identified by their initials, some of Roosevelt's early programs included the CCC (Civil Conservation Corps), CWA (Civil Works Administration), and the NIRA (National Industrial Recovery Act). These programs put millions back to work at least

A graphic from American Roofer *shows the 30 percent drop in the number of U.S. roofers between 1929 and 1931.*

temporarily, planting trees, constructing public buildings, and improving roads and drainage systems across the country.

The Roosevelt Administration's New Deal gave hope to millions of Americans unsure of the future. It also promised to nationalize business and industrial relations. The NRA (National Recovery Administration) was the cornerstone of Roosevelt's business program. The blue eagle emblem of the agency became a symbol of a new order to some and the badge of excessive government interference to others.

The most important part of NRA was its industry codes. These were industry-wide agreements on production quotas and price restrictions. Normally, such collusion over business volume would have met with anti-trust suits. But under NRA, industries gained the legal right to create such agreements in exchange for their willingness to allow workers to organize unions and participate in collective bargaining. Where workers declined to form unions, employers had to agree to pay a newly created minimum wage and adhere to strict weekly work limits. The NRA was supposed to produce a centrally planned, efficient economy.

The roofing industry received its introduction to NRA in June 1933. Talks were held in Washington with leading manufacturers, contractors and association leaders. There was a great deal of confusion, some optimism and much apprehension among the roofing contractors as the talks dragged on for almost a year. Finally, on May 10, 1934, the "Code of Fair Competition of the Roofing and Sheet Metal Contracting Division" was approved. The country was divided into local administrative committees that were to oversee the implementation of the Code. But the NRA brought only chaos to the economy at large and was floundering before the roofing code was even approved. It brought considerable worry to contractors already under considerable pressure. Few people in the industry were sorry when the Supreme Court declared NRA unconstitutional in 1935, and the codes were swept away.[26]

NRA was most successful in those industries that had strong trade asssociations. The United Roofing Contractors Association did not fit that description in the early 1930s. Its ranks had been thinned by the Depression. The loss of members sapped the Association of much of its effectiveness. In the depths of the Depression, the dark years between 1930 and 1933, two men struggled to keep the Association alive: J. Boyd Griffiths and Joseph A. Piper. They each served two successive terms as president of URCA during those difficult years. Piper, a Greenville, North Carolina contractor, had the particularly thankless task of having to contend with the NRA code negotiations.[27]

The United Roofing Contractors Association actually came very close to being a casualty of the Depression. The death in 1934 of Eugene Pope robbed the national office of an important source of continuity and leadership. Pope not only edited the Association's official magazine, but had served as secretary of URCA since its founding. Harriet B. Wagner took over the tasks of secretary, but the Association continued to lose vigor.

At the same time, a powerful new regional association was born that nearly overshadowed URCA. In 1935, roofing contractors in the Eastern states banded together to form the Northeastern Roofing, Siding and Insulation Contractors Association. NERSICA grew quickly, attracting members from across the country. The first executive director of the growing Association was a New York roofing contractor, James McCawley. To give his organization an effective mouth-piece, McCawley contracted with Sylvan Hoffman, a New York publisher, to put out a new magazine called *Modern Roofing*. NERSICA received a further boost when URCA, unable to find an effective editor for its journal, sold the publication to Hoffman. McCawley then became editor of the combined *American Roofer and Modern Roofing*. Without a publication, URCA's coffers and membership list continued to decrease.[28]

Tom Daly, who organized a small group of friends and past presidents to stave off NRCA bankruptcy in 1936.

Despite the heroic efforts of Piper and Griffiths, the Association again faced bankruptcy in 1936. Having invested much time and money in working on the NRA code, President D. A. Jackson resigned in frustration. Panicking, Harriet Wagner called Tom Daly, then vice president, and appealed to him for help. Never having been one to back away from a challenge, Daly picked up the telephone and called several of his good friends. He spoke with former presidents Piper, Griffiths and John Hession. They, along with Daly and his friend Irv Langer, each loaned the Association $200. The influx of capital enabled the URCA to continue paying Mrs. Wagner her salary and the rent for the office.[29]

Daly next turned his attention to the annual convention. Unfortunately, Jackson had done precious little planning for the 1937 affair, so one was hastily arranged. Planned by Langer, it was held in the Pfister Hotel, Milwaukee, his home town.

James McCawley, named executive secretary of URCA in 1938 and credited with playing a major role in reviving the Association.

Conventioneers, showing appreciation for the cool thinking that had again saved the Association, had the good sense to elect Daly president for the coming year.[30]

Oddly enough, it was McCawley, NERSICA's aggressive leader, who eventually revived the flagging URCA. Indeed, he went on to become one of the most influential men in the roofing industry.

James McCawley was born in 1899 in Scotland. At the age of 24, he immigrated to Canada. There he worked as a lumberjack, cutting trees with a crosscut saw and living in wilderness bunkhouses remote from any trace of comfort or civilization. Moving across Canada, he also found work in the wide open spaces of the western wheatlands. Neither of these jobs promised more than low pay and an aching back, so he kept moving, eventually crossing over to the United States. He took night classes at New York University and worked as a roofer by day. His academic background gave him an interest in journalism, while the roofing trade gave him the skills to found his own business.

He was a man with remarkable energy. While serving as executive secretary of NERSICA, he continued to operate his own contracting business and was in the process of writing a book on roofing. The book, *Roofing: Estimating, Applying, Repairing*, became a classic manual for young contractors. In 1938, thoroughly enjoying his job with NERSICA and as an editor, McCawley went to St. Louis to cover the United Roofing Contractors Association Convention.[31]

The annual convention continued to be a big event but organizationally, the URCA had hit rock bottom. The efforts of Daly and Langer had stabilized the situation, but memberships were not renewed nor were new members attracted. One source reports that by the time of the 1938 convention, only 10 supportive members remained. Those roofers who cared about URCA knew that this might be their last chance to save the organization. Former President J. Boyd Griffiths wanted a complete reorganization. He hoped to make a virtue of the small membership of URCA and transform it into an exclusive organization that would offer membership to only the elite of contractors. Most wanted to give an industry-wide association another chance. But clearly, if it was to have any future at all, leadership was required.

This fact was not lost on President Daly and Irv Langer. The 1938 convention was held at the Coronado Hotel in St. Louis. The lights in one of its rooms burned into the early hours of the February morning as Daly and Langer pondered the Association's future. Clearly, it did not lie with Mrs. Wagner, so they decided to fire her. The more difficult part of the task was deciding with whom to replace her. Knowing that McCawley was at the convention, and noting the friendship that had sprung up with his fellow Irishman, Daly suggested him for the post. Langer agreed, and at 4 a.m., the two men knocked at McCawley's hotel room door. Sitting on the end of his bed, the two men explained that all URCA could offer was its once-impressive reputation and the dedicated labor of a handful of concerned contractors. Yet, for whatever reason, McCawley accepted the challenge of reviving the Association and took the job.[32] With personal funds and borrowed money, he founded a new journal for the Association, *National Roofer*. The publication served as a vehicle for publicizing the revival of URCA and a way to pay McCawley a salary. Past presidents Griffiths and Piper assisted in the revival, as well as Chicago contractors C. L. Cockrell, Clyde Scott, Ben Esko, Bill Pennock, and Myron Powell; St. Louis roofing contractors John Reuter and Art Biebel, and many others across the country.[33] The convention adjourned after members elected Irv Langer to succeed Daly as the Association's president.

By the mid 1930s, the American economy also began to recover. Construction expenditures slowly began to recover from bottoming out in 1933. Once again, the number of new businesses exceeded the number of firms going out of business. In 1933, 13,000 more companies entered business than left. By 1934, 60,000 more entered business than left, and, after it dropped to just 1,000 in 1935, the number rose to 26,000 in 1936.[34]

The improvement in the economy had an immediate impact on roofing supply manufacturers. Between 1933 and 1935, the total value of roofing supplies produced increased by a dramatic 70 percent. Roll roofing increased by 10 million squares. Asphalt shingles production jumped by 3 million squares. Roof coating sales tripled, although concrete roofing tiles still lagged behind their 1929 production level. Metal and slate roofing materials did not rebound as quickly. In 1929, not a great year in itself, more than 5 million squares of roofing slate were applied. Yet in 1935, this figure was only 2 million. In fact, instead of participating in the general recovery, the Depression kept worsening for slate roofing. By 1936, the number of slate roofing squares applied dropped to a mere 1.1 million. But throughout the industry, there was reason for optimism. By the end of 1935, roofing factories were employing more workers than before 1929.[35]

Warren-Ehret, M. W. Powell and thousands of other roofing companies returned to profitability in 1936. Judging from the dividends authorized, Candler returned to profitability in 1935.

J. D. Candler emerged from the Depression aggressively pursuing the market. The company decided to provide an advertising budget equal to 5 percent of sales. The advertising was to be placed in newspapers, on radio and on billboards. The Board of Directors specifically wanted new work estimates to keep going out, thus insuring a steady flow of business. They hired a salesman who would do nothing but call on factory maintenance men "for the purpose of obtaining additional commercial business . . . " They hired another salesman who would be responsible exclusively for asphalt shingle roof sales. In 1936, the Board of Directors decided to establish a relationship with Chabot Hardware Stores, whereby Chabot would become a "sales contact for roofing prospects." They were to "be provided with advertising samples of asphalt shingles and a sign indicating their agency for J. D. Candler Roofing Company, Inc."[36] These efforts paid off handsomely by 1935, and the Company authorized a 10 percent bonus for its employees, or $2,255 from the year's net profits. A bonus based on 10 percent of an employee's wages was again paid in the first half of 1936 and at year's end bonuses of $4,500 were paid to salesmen, foremen and "key employees," with an additional $25,000 being set aside for bonus payments to hourly workers. Dividends of $6,000 were paid to J. D. Candler and his brothers Clarence and William.[37] Indeed, J. D. Candler Roofing had emerged from the Depression at full speed. By contrast, Warren-Ehret did not return its employees to full pre-Depression wages until 1939.[38]

The United Roofing Contractors Association advised its members to broaden their product lines. Many roofing contractors had intuitively been pursuing such a policy. If their principal line had been built-up roofs, they also began to offer asphalt shingles or sometimes even tile or slate roofs—anything to increase sales. The Kaw Roofing Company, headed by Thomas Daly, followed this policy. Through the first five years of the Depression, his firm enjoyed considerable and consistent growth. Daly, with a touch of blarney, was fond of telling competitors that his success stemmed from the early teachings of his Irish mother who counseled, "Be truthful and sincere at all times." His ability to sell every type of roofing system, as well as insulation and paints, allowed him the opportunity to bid on every construction contract in Kansas City.[39]

The trade journal *American Roofer and Modern Roofing* offered contractors advice on how to advertise their businesses by radio or billboards. Some contractors did not need advice on advertising. T. R. Heine, an Illinois contractor, had a regular radio spot that he used to pitch his roofing business in between solos played on his accordian.[40]

During this period, Nicholson and Galloway, the New York City contractor, also began aggressively advertising its services. Between 1931 and 1943, the company regularly ran ads in the *New York Herald Tribune* that were headlined by a catch phrase and illustrated. Nicholson and Galloway's product line was reasonably diversified and the ads eventually focused on most of its products. The sales pitches were direct: "We Cure Damp Walls," "We Stop Roof Leaks," "Stone Preservation," "Stop Corrosion," "Half a Million Square Feet of Leaking Walls Made Water-tight," and "North and East Exposures Are Vulnerable." The ads also asked "Is Your Plant Production Interrupted?"[41]

Time and Roosevelt's New Deal had made significant progress in getting the country back to work. By 1937, Roosevelt sensed that such an extensive array of programs might no longer be needed, and to minimize government expenditures, some of the programs were dismantled. He also increased taxes. Unfortunately, the country had not recovered enough for this action and more economic hard times followed, though not nearly as severe as those in 1930-1933.

Construction expenditures paused briefly on their post-1933 climb, falling off $100 million in 1938. Reflecting this dip, M. W. Powell's 1937 profit and 6 percent dividend fell to a net loss in 1938, with a small dividend being paid out of its surplus account.[42]

Continuing their climb, construction expenditures rebounded from 1938 and reached a post-Depression high of $9 billion in 1940. Powell recorded profits in 1939 and 1940 that allowed it to pay 10 percent dividends in each of those years.[43] Warren-Ehret was also profitable at the end of the decade, recording a 1.6 percent of gross sales profit in 1939, 1.5 percent in 1940 and 2.8 percent in 1941.[44]

J. D. Candler entered the 1938 depression just as aggressively as it came out of the one in 1933. Because of the uncertainty of its duration, "the managing director advised that expenses were cut wherever possible and that none would be incurred until such time as there appeared to be a reasonable expectancy for increased sales." Expense reduction included eliminating excess insurance, while adjusting the overhead rate to allow lower price bids to "meet the present low competitive bid," and modifying the estimate preparation procedure itself.[45] To encourage salesmen, commissions of 15 percent were paid on solicited sales, 13 percent on sales originating from telephone inquiries and 10 percent on sales to real estate management companies and banks. Candler's product line was further expanded to include asphalt brick siding for houses.[46]

Siding installation was the "new frontier" of the roofing industry during the 1930s. Clapboard had been a popular siding for frame houses since the Colonial period. But it required frequent painting and careful maintenance. During the last decade of the 19th century, an architectural movement called the "Shingle Style" used wood shingles to side the outer walls of its handsome suburban homes. But the movement was limited to the New England region and did not catch on elsewhere.[47]

By 1910, roll roofing manufacturers began to promote their product as a side wall covering. Although this system enjoyed a brief flurry of success, it did not meet with the general approval of builders and it soon disappeared. Throughout the 1920s, attempts were made to successfully apply asphalt shingles to side walls. Unfortunately, the shingles had a tendency to "wing out" after several months of exposure to weather.

But a breakthrough came about 1927. A young man employed by a local utility in western Pennsylvania to sell electricity to homeowners was surprised to find many steelworkers less interested in wiring than other home repairs. In particular, the soot from the coke ovens rendered all but the most recently painted homes gray and drab. The young man believed shingles could alleviate this and he set about solving the problem of holding the asphalt tightly to the wall. His solution was to nail the butt of the shingle with a small copper nail; it was simple but effective. In 1929, the first asphalt shingles made especially for side walls appeared. They were designed to look like bricks, an important selling point. Brick was a construction medium generally reserved for the homes of the affluent. Asphalt siding gave workers in frame cottages an inexpensive means of upgrading the appearance of their homes.[48]

The development of asphalt siding could not have been better timed. When the Depression brought on a sluggish roofing market, siding offered contractors a new field with growth potential. As one roofing contractor described it:

" . . . certain roofing contractors . . . simply refused to be put out of business by Old Man Rip Van Winkle. They just decided that if they couldn't do business one way, they would do it another way, so they turned to siding framed houses as an outlet for surplus energy. And now if Rip wakes up and comes back to town, he will find a new industry in the hands of the roofing contractors."

Improvements continued to be made in the siding field throughout the '30s. Asbestos was introduced to siding materials and the insurance industry further aided the boom by rewarding homeowners who had siding installed with lower rates. By the late 1930s, asphalt or asbestos-cement siding in roll form was introduced, greatly reducing installation time. By 1938, asphalt siding production had reached 1 million squares per year.[49]

The roofing industry seems to have enjoyed a love-hate relationship with FDR's New Deal. There were aspects to the Roosevelt program that were regarded as anti-business. The establishment of the National Labor Relations Board in 1935 increased the trend toward the unionization of workers. Union workers also were able to make a significant improvement of their lot during the New Deal. Union roofers worked shorter hours and received more pay than non-union workers during this period. Betweeen 1929 and 1939, the wages of unionized composition workers increased by 15 percent, while their hours fell by 8 percent. Organized slate and tile workers also gained by netting a 4 percent pay increase and an 8 percent hours cut in the same time span. Pay increases were largest in the big cities of the Midwest and Northeast. Workers in the South or rural areas lagged behind. The entire building industry experienced union pay raises, with roofers registering the fourth largest increases. However, the average hourly wage of a journeyman composition roofer was still the lowest in the trade.[50]

The Social Security Act in 1935 increased the burden of paperwork for contractors and the cost of doing business. A small number of contractors actually tried to dodge the provision

The Jack Johnson Company advertises its three-day special in 1932: 100 square feet of roofing applied for $4.75.

of payroll taxes for firms with more than eight employees by listing workers as partners. One New York roofing business actually claimed 18 partners. At the end of each job, the partnership was dissolved, only to be reformulated for the next project.[51]

The industry's reaction to the WPA was typical of roofing's attitude toward the New Deal. The Works Progress Administration, a federal jobs program, spent $11 million on small-scale construction projects between 1935 and 1941. Initially, roofers were pleased with the WPA. In its first year, it provided employment for 1,300 out-of-work roofers. The agency kept these men off the dole and at work on jobs that allowed them to maintain skills the industry would need after it recovered.[52] Yet within a few years, roofers were contending that the agency had outlived its usefulness and that roofs laid by WPA roofers were an unfair form of competition with private industry.[53] Although no industry publications openly attacked President Roosevelt, Eugene Pope, the editor of *American Roofer*, went so far as to urge all of its readers to vote in the 1936 election because of a firm conviction that "a reduced cost of government is vital to the interests of the industry." The same editorial noted that when Roosevelt was first elected, only 43 percent of the eligible voters participated in the election.[54]

On the other hand, the roofing industry heartily approved of the Federal Housing Administration (FHA). The agency's ambitious goal was to replace the blight of urban slums with new, low-rent housing.[55] The National Housing Act of 1938 was hailed as the instrument that would end the Depression for roofing contractors. Residential building had lagged behind the recovery of industrial and commercial construction. Yet residential building provided roofing with its largest source of business. The Act directed federal dollars to this problem. The FHA agreed to guarantee mortgages up to 90 percent for homes costing less than $5,000. The Act also funnelled support for construction of multi-unit projects. Most important of all were the loans offered to encourage home repairs and modernization. The *American Roofer* estimated that there were 2 million homes in America that needed reroofing, but whose owners had delayed it because of the Depression. Flintkote Company got one up on its competition by releasing two sales plans on the very day Roosevelt signed the Housing Act. These plans guided contractors who wished to profit from the new legislation and outlined steps to turn its provisions into customers. In six months, more than 140,000 notes had been guaranteed by the FHA, totaling $67.9 million. Roofing contractors had snared 9 percent of all projects initiated under the Act.[56]

FHA programs proved especially important to roofing contractors in the Southern and Western states. In one month, the Tennessee branch of the American Asphalt Roofing Company captured more than $300,000 in sales financed by FHA. The Copper Roofs Company of Phoenix was able to expand its staff from five employees to 46 within one year of the FHA program's start-up. The Mountain States Roofing Company of Denver reported a 300 percent increase in sales after a year of FHA programs in that area.[57]

There was one minor incident that directed the United Roofing Contractors Association's ire against President Roosevelt. In 1937, Roosevelt had considerable trouble getting his legislation through Congress. It seemed the economy was recovering and with the nation no longer in a state of emergency, even Democratic members of the House were in an independent frame of mind. To generate public support for some of his pet domestic programs, Roosevelt decided to take a few jabs at Congress during one of his "fireside chats." What made those regular radio broadcasts

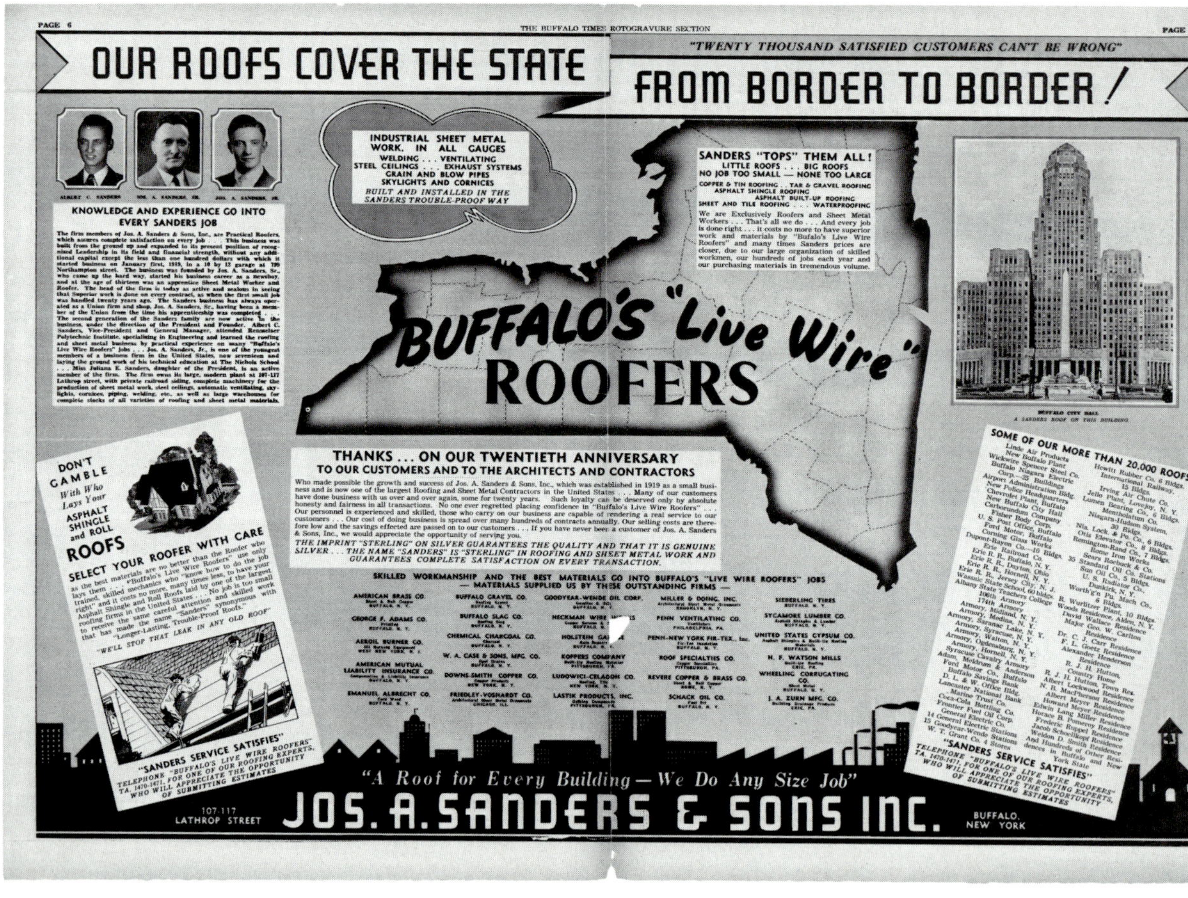

Jos. A. Sanders & Sons celebrated its 20th anniversary in 1936 with an impressive spread in The Buffalo Times *rotogravure section, despite the grim state of the economy.*

so successful was FDR's ability to explain complex issues through the use of common, everyday expressions. Unfortunately, in this particular chat, he used the theme "The time to repair the roof is when the sun is shining" to explain his desire for congressional action. With the fall construction season ending in the northern states, this remark could not have been more poorly timed for roofing contractors. URCA had long waged a campaign to promote the perception of roofing as a year-round industry. Charles Fitzpatrick, a roofing contractor, sent an open letter to the president, suggesting that fair-weather-only roofing was an idea technology had relegated "to the limbo of outdated things." Although Fitzpatrick took a few friendly swipes at the president, he used the letter largely to remind the general public and "the few surviving hibernating roofers" that they need not wait until the sun shines to have roofs repaired or applied.[58]

By the end of 1938, roofing contractors across the nation could clearly see the end of hard times. Census returns for that year reported a "dizzying upward climb in the volume of built-up roofing." In the two years between 1935 and 1937 alone, there was a 79 percent increase in the volume of tar and asphalt felts sold. Metal roofing and asphalt shingles showed similar increases.[59] Although 1939 brought a return to prosperity for contractors, there were indications that further disruption of the economy lay ahead. Europe was once more at war. As armies mobilized across the Atlantic, Americans debated their own prospects for continued peace and prosperity.

The lack of available manpower at a time when the industrial production rate was 100 percent created an opportunity for women to enter the ranks of workers. Rosie the Roofer came into her own.

CHAPTER 7

Challenges and opportunities: World War II roofing

"Democracy is based upon the conviction that there are extraordinary possibilities in ordinary people." HARRY EMERSON FOSDICK

In the fall of 1939, the primary question among contractors was, "How will the war affect roofing?" Comparisons were drawn between 1914 and 1939; prognostications were offered. But the general concensus of American roofing contractors was that the war would lead to an "upsurge in business activity" in the United States. Optimism ran high among contractors. The Detroit contractor A. J. Tripp of the Veteran Roofing Company reported that residential reroofing jobs in Detroit doubled within a month of the outbreak of war in Europe.

Naturally, the reason for the sudden surge was the general increase in American industry to supply the warring nations with non-military goods. Herbert Abraham, president of Ruberoid Corporation, confidently predicted: "Additional housing facilities for augmented working forces would probably be required in various industrial centers . . . " But most leaders of the roofing industry also agreed with the cautious reminder of Joseph Stelwagon of Philadelphia's Stelwagon Manufacturing Company: "I feel that as long as we remain out of the war, the building of residential and manufacturing plants should increase. However, should the United States enter the war, building would, of course, decline."[1]

Roofing contractors did not want to think of a decline. They had just completed the long climb out of the Depression and were interested in expanding their businesses. Besides, they were being hard-pressed by competition in the building trades. One of the most aggravating forms of competition came from lumber dealers. It had long been the practice of a small number of lumber dealers to team up with local carpenters or other tradesmen and bid on roofing jobs. Supported by a retail reputation and close association with roofing supply manufacturers, the dealers had an unfair competitive edge over the independent contractors. As the building trades recovered in the late 1930s and early 1940s, lumber dealers increased their efforts to edge out the local contractors. In Massachusetts, an organization called the Cooperative Lumber Leaders took particularly aggressive action against independent contractors. The organization began an ad campaign attacking contractors as unethical and inefficient. The ad contended that:

"Roofing canvassers and salesmen, who ring doorbells in order to locate customers, are unscrupulous. Their arguments are false, their facts unreliable. If the homeowner does sign on the dotted line, he will find he has driven a costly bargain which he will regret. For honest, reliable roofing, siding or insulation services, the homeowner should patronize the lumber dealer."

Local contractors were stunned and spurred to action by this and other attacks by the lumber dealers. But, although they found the individual dealerships apologetic, the lumbermen collectively refused to retract their statements.[2]

To a certain extent, the roofing business had no one but itself to blame for the criticism. Even the United Roofing Contractors Association admitted that roofing was plagued by irresponsible, fly-by-night elements. As the pace of construction quickened, the ranks of these unreliable roofers increased. The whole question of ethics and regulation became intertwined with the individual contractor's battle with competition from local lumber dealers and inexperienced "jobbers." The June 1939 issue of *Collier's* proclaimed, "Either the construction industry is going to fix itself or the government is going to step in and fix it."

Legislation was introduced into the House of Representatives to license contractors. The time had come, many contractors felt, for roofing to do something to improve its image as an industry. One manufacturer argued that roofing had arrived at the crossroads and urged action at a national level.[3]

The founding of the Sheet Metal, Roofing, Heating and Air Conditioning Contractors' National Association in September 1939 was one attempt to address this problem. The idea behind the new Association was to combine the strengths of several allied building trades.[4] Unfortunately, before the Association could become an effective force, the United States was at war.

The Japanese attack on Pearl Harbor energized and united America. Military build-up and industrial expansion, which had begun modestly in 1939, were now accelerated to capacity. Unemployment ceased to be an issue as draft boards and factory employment offices competed for manpower. The recovery agencies of the New Deal were succeeded by new alphabet agencies created

Roofing crews work furiously to finish a bomber plant during the war. The federal Supplies Priorities and Allocation Board made sure the war industries received all the materials they needed.

to direct and sustain the war effort. The WRB (War Resources Board) was created to convert the nation's industry from peacetime production to wartime production.

Such controls were clearly necessary. After war broke out in Europe in 1939, one of the first reactions in the building industry, including roofing, was a wave of "hysteria and speculation." Herbert Abraham of Ruberoid Corporation observed:

"First, there has been some hysterical buying in an effort to build up, at current prices, inventory stocks for use by the purchaser in the regular course of his business; and, second, there has been, in all probability, some purely speculative buying, motivated by the expectation of being able to resell in the near future at a substantial profit."

To control prices, wages, and overall production, three other important agencies were created: The SPAB (Supplies Priorities and Allocation Board) was charged with making certain that war industries got all the raw materials they needed. The OPA (Office of Price Administration) was created to oversee wage/price fluctuations so as not to induce runaway inflation. The OWM (Office of War Mobilization) maintained a watch over the entire wartime economy. Finally, the NWLB (National Wage Labor Board) was to assure continuous war material production by mediating labor disputes, as well as making sure wage requests were kept within the 15 percent of pre-war wage guidelines established by the government.[6]

Successful though these agencies were in helping the war effort, they did not enjoy universal acceptance. Labor unions, for instance, pledged not to strike during the war in return for NWLB. But they were not happy with NWLB because it stabilized wages at a time when prices

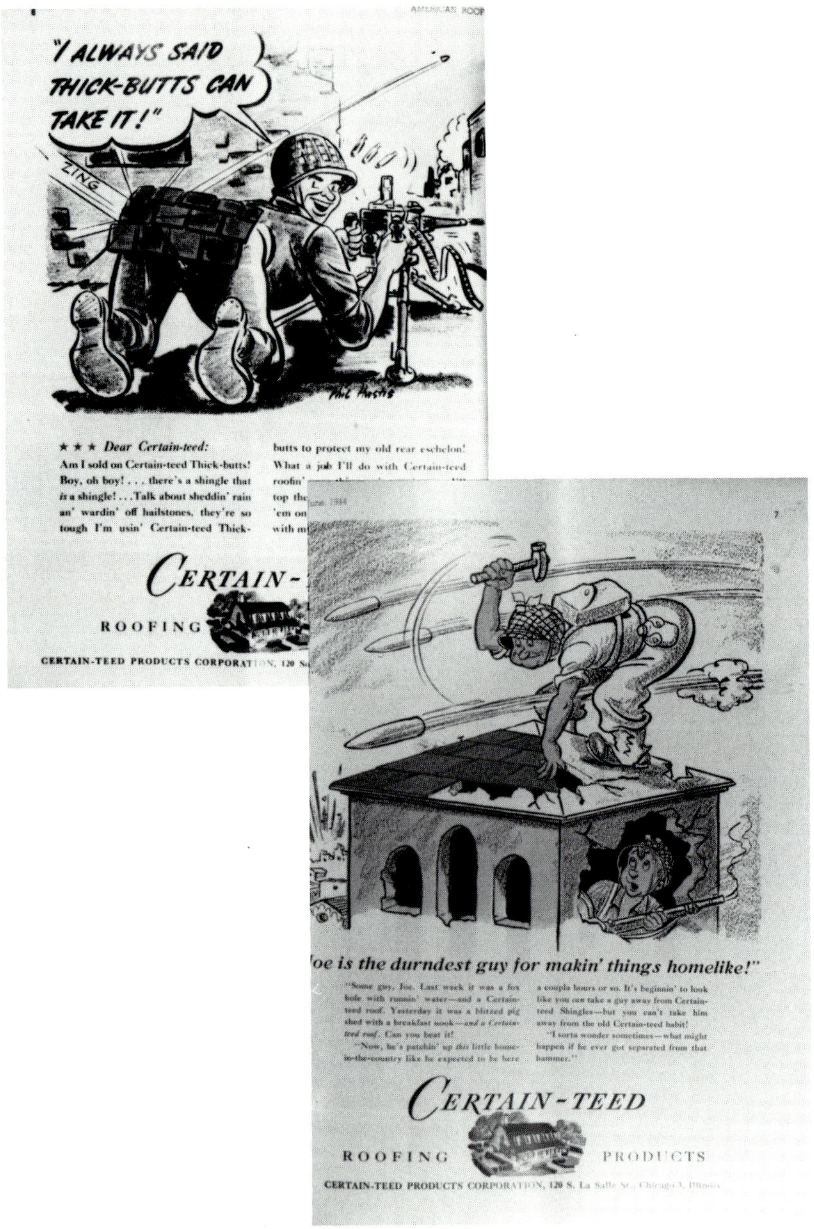

were up 26 percent. In addition, workers had a minimum 20 percent withheld from their checks for taxes and still were expected to invest 10 percent of their wages in war bonds.[7] The Executive Council of the roofers' union took direct aim at the OPA and the 26 percent rise in prices. A stinging condemnation resulted, that read in part:

"The price control program with regard to foodstuffs has broken down. The feeble, fumbling, half-hearted efforts of the Office of Price Administration to keep food prices in check has resulted in almost complete failure. Despite repeated promises, there has been no sincere and effective attempt to roll back prices to a reasonable level. Instead, the Office of Price Administration has created confusion and chaos in the nation's food markets by a continuous series of complicated and contradictory rulings which have made a mockery of price control."[8]

The Executive Council concluded by stating that it was "convinced that a simple and fixed system of price ceilings is required to eliminate profiteering and to facilitate enforcement. This program is so clear-cut as to assure complete public

CertainTeed Roofing Products launched a series of war-related advertisements. Left: "I'm usin' Certain-Teed Thick-Butts to protect my old rear echelon!" Right: "You can take a guy away from Certain-Teed Shingles—but you can't take him away from the old Certain-Teed habit!"

understanding and universal compliance."[9]

In 1941, there were 1.5 million people in the country's armed services and by 1945, that number had swelled to more than 15 million. The resulting shortage of men in the country, at a time when industrial production was at full capacity, opened the door for middle-aged women to enter the industrial workplace. There were the equivalents of "Rosie the Riveter" in the roofing industry, as there were throughout the construction business. Some wives took to the roofs to aid their husbands, while other contractors started hiring women as apprentice roofers. H. F. Stuart of the Dutchess Home Improvement Company in Poughkeepsie, New York, used two women on his crew. He later recalled, "It was a great sight to see how nicely the ladies could shingle."[10] A Florida contractor tried using women on the roof, but thought they did not work as effectively as men. But this misanthrope was in the minority.

Ruberoid Corporation roofers in Britain used female applicators. The first week the female crew accomplished 10 percent more than an all-male crew. By their second week on the job, women were laying 25 percent more roofing surface than the men. After three weeks' work, the superintendent was approached by one of the women. She complained, "Can't you put those men in some other job? They're only in our way."[11] Other contractors brought women into the office or onto their sales staffs. A Midwestern contractor employed housewives to go door to door with literature on reroofing.

The labor shortage permeated the roofing industry during the war. J. D. Candler of Detroit entered 1940 still marketing its services and products aggressively on the radio and in display newspaper ads. Business continued to be good well into 1942 as Candler continued to pay dividends to stockholders and bonuses to employees. By the end of 1942, however, a conscious decision was made not to advertise. Business had increased beyond what the crews could handle. The problem was that the war created a labor shortage. The problem grew so acute in 1943 that Candler began trying to secure draft deferments for some of its employees, without success. The problem continued to worsen, and by 1945, Candler was using caution in bidding large contracts, electing to concentrate most of its depleted resources on repair, maintenance and small, new jobs.[12] Lawson Roofing of San Francisco provided an extreme example of the way the war hurt some contractors. The company was only able to maintain one crew, and thus elected to pursue small residential and commercial repair jobs.[13]

The overall severity of manpower shortages varied from contractor to contractor. The War

Manpower Commission listed "Roofing" as well as "Building Alteration, Maintenance and Repair" as an essential activity. Unfortunately, many roofing contractors were not aware of the significance of this ruling. Local draft boards were instructed to selectively enlist workers from these industries and only "to the extent and in such numbers as may be required to meet the minimum essential needs of the community." Also, roofing mechanics were required to receive their employers' permission if they wanted to seek work in other essential war industries. A Minnesota roofing contractor was able to convince the government to order three of his men out of a steel plant because of this regulation. But ignorance of the regulations hurt many other contractors. An unwillingness to match the higher wages of industrial jobs also hurt contractors.[14]

Labor was not the roofing industry's only constraint during the war. The market for residential roofing was constrained by wartime credit management. The infamous "Regulation W" restricted loans for the repair and maintenance of private property. Although an estimated one in 15 homes was ready for reroofing each year, federal regulations forced ethical roofers to demand cash sales from their customers. This was a burden for the average worker, who was paid less than $2,000 a year. Even small reroofing jobs of $150 became a burden, with no chance of financing. It was also difficult to reach potential customers during the hectic years of the war. Contractors who still had sales staffs found canvassing neighborhoods difficult, because so many war workers were operating on staggered shifts.[15]

Manufacturers found it difficult to deliver roofing supplies during the war. New industrial construction was limited by the need to devote all resources to war-essential industries. There was little opportunity to build or expand roofing product factories. But more than 60 percent of the industry's normal production went for reroofing. Although some reroofing could be delayed, most was needed—war or no war. The strain was felt particularly by the asphalt roofing producers. Before the war, asphalt roofing accounted for about 70 percent of the roofing market. Yet, because of war restrictions on metal and wood, federal planners called for an increase of asphalt and tar roofing to 90 percent of the market.[16] This sort of expansion without construction was only possible because of efficiency. The War Production Board ordered manufacturers to streamline their production to certain industry-wide specifications.

For the asphalt roofing and siding business, this regulation brought a marvelous transformation. Before the war, there were 306 different products on the market. Early in 1942, this was cut to 133 products, and by December 1942, after only one year of war, the industry was producing only 20 different lines of asphalt roofing or siding. In spite of the fact that the regulations eliminated items that some manufacturers had gone to great pains to promote, most companies were in favor of the regulations in light of the limitations on manpower, physical plants, and materials.[17]

Not all actions by the War Production Board met with approval by roofing contractors. In July 1942, all roofing and sheet metal businesses were requested to submit an inventory of "idle and excess" copper and brass stock. Many contractors were then forced to sell these materials to the WPB's Copper Recovery Corporation. Some contractors suffered large losses. There were cases of roofing businesses being paid only 30 percent of the value for their reserves. The URCA was outraged. It described the program as "Hitlerism at work." Roofer W. L. Wright complained, "To the Nazi mind, it may be agreeable to have the Gestapo drive up in a truck and confiscate copper cooking utensils without even a by-your-leave, 'because the Fuhrer needs it' . . . but WE live in the United States!"[18] Metal roofers were given more grief when the War Production Board set limits on the amount of tin in solder. The limitation greatly weakened the alloy's effectiveness. Contractors from around the country complained to URCA, which eventually was able to convince the War Production Board to increase the tin content in solder by 50 percent.[19] Like so many other businessmen, roofing contractors were discovering that cooperation was required to compete in the highly regulated wartime environment.

Myron W. Powell, who served as president of URCA in 1941.

The roofing industry's most important wartime problem came with United States Army contracting procedures. Even before Pearl Harbor, the Army financed a major construction program to expand its bases and test facilities. But roofing contractors seldom benefited from this boom. In its desire for speed, the Army Quartermaster Corps elected to have its general contractors apply all roofs. The United Roofing Contractors Association and roofing contractors around the country were outraged. James McCawley, in a complaint to the War Department, stated: "Some of our members are under the impression that the Quartermaster Corps, during the past year, has been trying to run them out of business." The Army agreed that general contractors, even if using experienced roofing crews, did not apply roofs as well as professional contractors. Yet, instead of involving roofers as subcontractors, the Army decided to have them called in after construction was completed to repair the roofs put on by general contractors. Painters, sheet metal workers, and other craftsmen were asked to swallow the same bitter pill. Aside from the waste of resources and the maintenance problems caused by this approach, roofing contractors were justly concerned over the precedent. If general contractors got used to handling roofing on their own during the war, might they continue to cut out the roofing contractor after the war?

To deal effectively with wartime procurement policies, manpower shortages, industrial production management, and price regulations, the roofing industry needed to take united action. That leadership came in part from URCA and in part from new organizations such as the Roofing Industry Advisory Committee, the asphalt roofing producers' National War Council, and local contractors' associations.

The critical leader of the United Roofing Contractors Association during the war years was Myron Powell. As a youth, he was more interested in poker than roofing and was in many ways an embarrassment to the business his grandfather founded. However, as he matured, Myron Powell became not only a competent contractor, but a major figure in the industry. He was born in 1892 and was a vigorous man of 48 when his colleagues elected him president of URCA in 1940. He was re-elected to that post the following year and such was his leadership that when his term as president expired, Powell continued to be URCA's principal troubleshooter. His first action was to

Housing projects for employees of war-related industries contributed to a surge in demand for construction.

take aim at the Army's exclusion of roofing subcontractors. But his official complaints as URCA president were rebuffed by the Army as so much sour grapes.

Not discouraged, Powell formed a National Roofing Industry Committee, with a handful of distinguished contractors from each part of the country. They complained about the Army policy, not on the grounds of its discrimination against roofing contractors, but because of its waste of tax dollars and valuable resources. With congressional committees keeping an ear cocked for waste and inefficiency, the Army Quartermaster Corps caved in to Powell's ploy. Roofing contractors were in a position to participate in defense projects. It was the most important victory of the war for the roofing industry.[20]

To help roofers take advantage of war work, URCA reorganized itself. The organization was divided into nine zones covering the various regions of the United States. A 10th zone embraced the entire dominion of Canada. The new URCA zones corresponded with Army Quartermaster Corps construction zones to ensure close cooperation between the Association and the contract officers.[21]

A new feature of URCA during the war was industry research. The Association sponsored investigations into subjects vital to the industry's

The promotional brochure for the URCA meeting in Chicago in 1940 proclaims the convention theme: "Away with the gloomy past! Forward with the Forties!"

war effort. Roofing specifications that might sustain thermite fire bomb damage were studied. Engineering studies were also conducted on roof coverings that might resist demolition bombs. For munitions plants, roofs were needed that would quickly disintegrate so that the force of an accidental explosion would be spent on the roof, not on the walls of the building. There were even studies of camouflage patterns for various roofing systems.[22] As part of this program, the United Roofing Contractors Association sent Executive Secretary James McCawley to Great Britain to study how the British roofing industry contributed to the war effort. During the 1943 trip, McCawley saw the emergency roofing work performed by British contractors in the midst of a blitz and studied British camouflaging techniques. McCawley also submitted a report to the United States Office of Civilian Defense and the Federal Fire Council on Roofing and American Civil Defense, a report which, fortunately, was never needed.[23]

The pace of roofing during the war was such that URCA could not carry the entire burden of industry leadership. Although James McCawley's tenure as executive secretary had revived the Association, it still could claim as its membership only a fraction of the more than 6,000 contractors in the United States. To deal with the wide range of issues it could not effectively address, URCA encouraged the establishment of local trade associations. Such organizations were seen as more effective tools for dealing with labor problems or unfair business competition. Like the Depression before it, the war had the effect of increasing the number of trade associations in the roofing industry.[24]

Manufacturers were important to the work of both URCA and many of the local associations. But in terms of protecting their own interests, manufacturers often followed a separate course. One of their early concerns was the price standards of the Office of Price Administration (OPA). Maximum prices were established for asphalt and tar roofing products. To gain input in raising those ceilings, executives from some of the leading manufacturers (Philip Cary, Flintkote, Johns-Manville, and others) lobbied the agency for an official advisory committee. In April 1942, a special Roofing Industry Advisory Committee was established and the manufacturers thereafter worked with OPA on industry problems.[25] Later that year, the asphalt roofing producers set up a National War Council. The Council was designed to serve as a liaison with the War Production Board. Its initial concern was how to efficiently supply needs for roofing manufacturers. But the National War Council also made plans for emergency repairs of bombed or shattered roofs in case of an enemy air

JOLIET HERALD-NEWS

VOL. XXXVII. No. 40—26 PAGES FRIDAY, DECEMBER 27, 1940 Three Cents

ADLER GETS $400,000 CONTRACT

Sea Raider Attacks Pacific Isle

Late News Flashes

SHIP WITH JAP FLAG DAMAGES BRITISH AREA

English Conduct Hunt for Enemy Vessels in Region

SOLONS SEE HOPE FOR NEGOTIATED EUROPEAN PEACE

Greeks, Italians Wage Fierce Fight in Klisura Area

JOLIET FIRM TO ROOF SHELL LOADING PLANT

2 Million Square Feet of Material Will Be Made by Ruberoid

Make Plans for Greek Relief Campaign

WOMAN HURT IN CAR COLLISION

Begin Changing Jail Yard Into Flower Garden

STUDY CIO PLAN TO PUSH DEFENSE

LE HAVRE HEAVILY DAMAGED BY R. A. F. BOMBING ATTACKS

BRITISH BOMBERS HAMMER AT NAZI "INVASION PORT"

Calendar Firm's Reports Indicate Record Season

Fall of Bardia Awaits British Go-Ahead Signal

"I See by the Pa

Street Comment

Chuckle Box

It's On Page . . . .

attack. The Council feared, "Hitler may launch such attacks on our major war-producing centers, as soon as our air power threatens to eclipse that of the Luftwaffe." Drawing upon thc experience of Eastern manufacturers, who in 1939 operated on a 24-hour basis in the wake of a New England hurricane, the Council laid plans to instantly supply "vast quantities of emergency roofing material, labor, and transportation."[26]

In spite of the high level of patriotism among most roofing contractors and the growth of trade organizations committed to quality roofing, the industry was plagued during the war by its old nemesis—irresponsible and unscrupulous contractors. Federal officials received numerous complaints of roofing and siding contractors exceeding price standards. In 1943, the Office of Price Administration cited many examples of unscrupulous contractors dodging federal regulations. One price regulation, Maintenance and Repair Order L-41, prevented contractors from accepting contracts over $200 (except in emergencies) without reporting to the agency. A Michigan siding con-

The big news in Joliet, Illinois in 1940 was the J. L. Adler Roofing Company's contract to roof the local shell loading plant. Two million feet of Ruberoid roofing and siding was to be used on the structure, which was thought to be "the largest single order ever placed in the industry."

tractor sought to avoid this on a job where materials and labor should have cost between $90 and $200. Instead, the contractor got away with charging $529 by giving the salesman a commission of $432! In October 1943, the War Production Board and the Office of Price Administration assembled 700 Detroit contractors and warned against that and other illegal practices. In several states, Better Business Bureaus discussed a formal code of ethics for the roofing industry. The Federal Housing Administration suggested that "the roofing contractor clean house." URCA called for action or "the laundering would be enforced by outside agencies."[27] However, the frantic nature of wartime construction pushed all issues to the back burner before long; the principal aim was winning the war. Although the ethics issue was allowed to be buried, it was not one that could be completely ignored, and it would fester until later addressed.

For many contractors, the war was a period of growth, not so much in the volume of their businesses, because often shortages constrained such expansion, but in their management of large-scale, short-term projects. They experimented with the use of labor-saving equipment. For an aircraft plant in Dallas, roofers were required to lay a composition roofing system over a 20-acre area. Operating at a breakneck pace, they averaged an acre a day. At one point the workers laid 17 square feet per second![28] It was not unusual for such big projects to have a team of roofing contractors working in concert. At a Fort Worth bomber assembly plant, a firm known as Consolidated Roofers applied the 4,000-foot-long roof. It was actually a combination of three of the Dallas-Fort Worth area's most prominent roofing businesses: Lydick Roofing Company, Hamilton Roofing Company, and Builders Material Company. Alone, none of these firms could marshal the labor or equipment to meet the Army's production schedules but, by cooperating, they were able to share in a large contract. Similar cooperative agreements helped keep roofing contractors busy in spite of wartime shortages.[29]

In 1945, allied military forces brought World War II to a victorious conclusion. Peace brought an end to 16 years of economic and political crises that stretched back to the beginning of the Depression in 1929. The roofing industry came out of the troubled era more efficient and better equipped. However, in terms of structure, it was not greatly altered. Of the 5,000 or so firms that specialized solely in roofing, fewer than 20 percent handled more than $25,000 of business. The low volume reflected the largely family-oriented structure of the industry. Most roofing companies, in war or peace, had only a handful of employees and were managed by one man, almost always an active proprietor. Because of their small size, they did not profit as much from wartime construction as the larger companies, nor were they open to the same regulation relief as firms that participated fully in war work.[30] During the Depression, their emphasis was on feeding their families and staying in business. Once they were at war, they grumbled as their mechanics and even their sons went off to fight, but they fell in step with the patriotic public sentiments of the era. New York contractor Alfred E. Neulander captured this when he called for small contractors to "Trowel-coat Hitler . . . under four courses of mud and sand . . . "[31] With the war ended, these men were tired of the sacrifices of crisis and eager for peaceful prosperity.

NRCA's Board of Directors meet at the Sherman Hotel in Chicago in 1952. Board members were paying $6 a night for a single room; $18 and up for suites.

CHAPTER 8

Coming of age: roofing during the Baby Boom

"The more extensive a man's knowledge of what has been done, the greater will be his power of knowing what to do." DISRAELI

When the guns fell silent in the summer of 1945, the United States was poised for a period of expansion never before experienced. After enjoying the prosperous 1920s, Americans had been forced to show tremendous restraint during the Depression of the 1930s and World War II. Now the Depression was over and the war was won. Consumer demand was ready to explode and Americans were eager to buy products as novel as television sets or as necessary as houses.

The construction industry was thought to be the vehicle for stimulating the economy and rebuilding the nation's housing. Building had fallen off dramatically during the Depression and was just starting to recover when the war curtailed it. Some construction took place during the war, but it was largely industrial. Housing projects were limited to the essential goal of providing shelter for war workers. A building revival would not only put people to work in construction, it would also employ many more to work in associated manufacturing fields.

The United Roofing Contractors Association anticipated the post-war construction boom. Choosing the theme "More Business for the Small Contractor" for its January 1945 annual convention, the Association tried to show small contractors how to take advantage of the business increases that would soon confront them. Sessions at the convention were designed to acquaint roofing contractors with the financial services available at banks, how to advertise and sell their products and services, and how to create a positive local identity.[1] The Association's help with these practical problems won it many friends. More than 100 members joined URCA between the 1944 and 1945 conventions.[2]

The war and postwar changes affected roofers in a variety of ways. For some, it was a period of tremendous prosperity. For others, the war was a blow to their businesses and an inducement to try other trades. That was the case of Milton Carpenter and W. R. Taylor.

Carpenter entered the war a young, highly successful contractor. In 1942, he was even elected president of the United Roofing Contractors Association. But his life changed suddenly when he received a notice from his draft board. He never finished his URCA term and, in a matter of weeks, he was boarding a train for boot camp. Carpenter completed his tour of duty as the war ended. But his homecoming was bitter; his business was a disaster. The company's bookkeeper had tried to keep it going with no success. Supplies had dwindled to nothing and half of the warehouse had been rented out. Carpenter decided to leave the roofing contracting business after deciding that the company could not be salvaged without tremendous effort; he was also presented with a chance to obtain a franchise for the Patent Scaffolding Company for Missouri, Kansas, Oklahoma and Arkansas. He subsequently spent several years in the scaffolding industry before deciding to make another career change. From his humble political beginnings as URCA president in 1942, Carpenter was ultimately elected treasurer of the state of Missouri, leaving that office in 1965.[3]

Another World War II contractor who chose to leave roofing was W. R. Taylor. Taylor began farming part time in 1943, probably as a result of the demand for foodstuffs created by the war. Finding work on rows of crops more to his liking than on rows of shingles, Taylor gave up roofing in August 1945, and became a full-time farmer.[4]

Some roofing contractors took advantage of the opportunity to do defense work and came out of the war stronger than ever. F. J. A. Christiansen of Milwaukee was one contracting firm that availed itself of this opportunity. It aggressively sought and obtained government work during

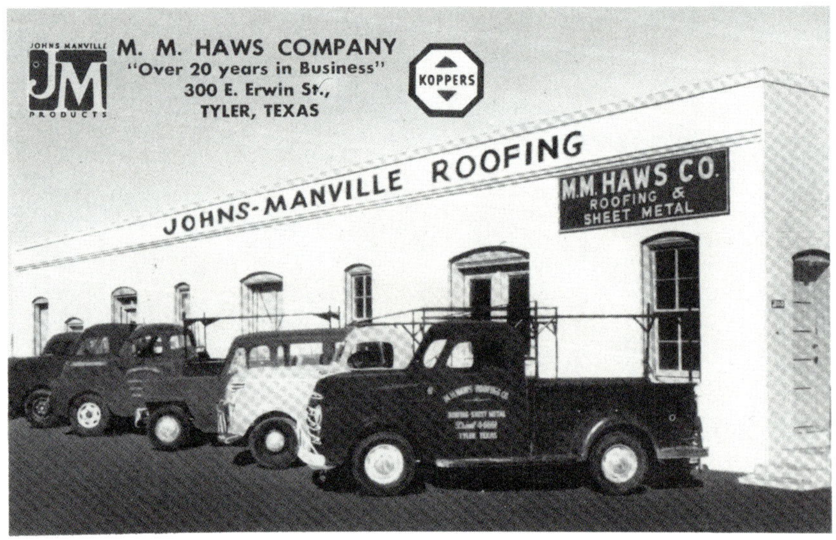

A 1953 example of a contractor's direct mail piece; the back of the postcard boasts the M. M. Haws Company's use of "world-famous Johns-Manville asbestile treatment."

the war. Christiansen worked on war plants and military bases in Kentucky, Indiana, Illinois and Wisconsin. The company had no problem getting men to work on these jobs; the government helped direct the labor supply to contractors engaged in defense work. The government also provided all the materials and equipment for this work. Although it is impossible to generalize about what this work meant to the economic health of the firms that participated in it, Christiansen's volume grew from $60,000 in 1932 to $800,000 in 1944, largely because of the government work it pursued.[5]

On November 1, 1945, President Truman called together various members of the construction industry to discuss their problems and plans, and to outline the strategy for accomplishing his goal of 20 million new homes built by 1956. James McCawley addressed the meeting; he advised the group that a major impediment to that objective was the shortage of skilled mechanics and apprentices. A significant result of the meeting was the creation of the Construction Industry Advisory Council, of which URCA was a member—the "... first time the URCA has secured [such] recognition on a national construction organization ..."[6]

What held the roofing industry back from the potential growth of the era were its wartime problems of manpower shortages and material scarcity. The lack of apprentices in particular was a major hindrance in accomplishing President Truman's goal. Simply to maintain the level of construction reached in 1940, the construction industry needed 249,000 apprentices in training at all times.

In the roofing industry, specifically, there were 32,700 skilled workers in 1940. To maintain that labor supply, there had to be 3,800 apprentices in training at all times. Of these, 1,400 had to be employed by roofing contractors each year, and 600 of those had to reach the journeyman level each year.[7] In 1940, construction expenditures had reached $80 million. That figure skyrocketed to $653 million in 1946 and to $733 million in 1950.[8] That meant that to meet the construction levels of 1950, the number of skilled mechanics, then apprentices, would have to increase greatly over the 1940 level.

URCA, in conjunction with the Association of Slate, Tile and Composition Roofers and the Federal Bureau of Education, created a roofing industry apprenticeship program that was designed to help meet the demands of the expanding construction industry. Designed to last three years, the program accepted people 18 or older and required them to split their time between roof work and classroom work. By the conclusion of the three-year period, 144 hours had to have been spent in the classroom. Apprentices in the first six months of the program were paid at 65 percent of the journeyman's rate. It increased 5 percent for each six months thereafter, finally reaching 90 percent during the last six months in the program.[9]

In addition to the lack of apprentices, the roofing industry also faced a critical felt shortage. Worse than the shortage of apprentices, the lack of felt could bring the industry to a standstill, thus forcing contractors out of business. In fact, the felt shortage directly threatened the industry's ability to meet the government's goal of building 3 million new homes by 1948. In 1946 alone, construction on 50,000 homes was delayed because of the felt shortage. Because it was an issue so important to the contractor's well being, it occupied center stage at the 1946 URCA convention in St. Louis.[10]

Felt was made of rags, and there was a worldwide rag shortage. Rags were imported from around the world, just to make roofing felt. Excellent sources for the fibers included satin garments, coats, vests, trousers, carpets, stockings, silk shirts and blouses. But there was a clothing shortage in Europe. Egypt, North Africa, China and Japan also became major rag importers. Around the world, people wore their old clothes longer instead of throwing them away. The rag shortage in the United States was intensified when people, to help the needy overseas, began sending their old garments to war-ravaged lands. Rag merchants could demand twice as much for rags in the world marketplace. Roofing contractors were caught between the international market and the producers of wiping rags, whose demands further boosted rag prices. Production of roofing felt was therefore slowed by forces beyond the control of private industry.[11]

Felt manufacturers were willing to pay more for the rags, and the rag merchants did not want to lose the roofing industry as a customer, but the prices were regulated by the war-time Office of Price Administration. There was nothing either could do.

Realizing the desperate situation, the URCA members at the St. Louis convention authorized a trip for a committee that would plead the contractor's case to the government. The resolution read:

"In view of the inescapable fact, now so fully recognized, that there is an extremely serious and rapidly growing shortage of the saturated roofing felts that are so essential to the application of built-up composition roofs . . . Be it resolved, that URCA send a committee to Washington, D.C. . . . : to discuss the imperative need for an increased supply of saturated roofing felts."[12]

This emergency measure reveals not only the importance of the felt shortage, but the increasingly significant role the federal government

An artist's depiction of the furious rate of construction during the post-World War II years.

played in the health of the roofing industry. Lobbying in Washington gradually became part of the URCA mission.

URCA President Charles Cockrell and Executive Secretary James McCawley went to Washington on April 15, 1946. They pleaded the contractors' case at the White House, the Civilian Production Administration, the National Housing Administration and the Office of Price Management.[13] These meetings helped remove government obstacles. The market was then free to sort out the felt shortage unrestrained by outmoded wartime regulations.

Gradually, as the manpower and supply problems were resolved, the industry grew as never before. In 1945, 1946 and 1947, 747,000 more contract construction firms entered business than left it. That trailed off to 203,000 for 1948, 1949 and 1950, but it indicates how quickly the construction industry came out of the starting gate at the war's end.[14] The number of roofers also grew dramatically during this period, from 33,000 in 1940 to 50,000 in 1950.[15]

J. D. Candler, sharing the good times, achieved a 1946 net profit of $30,000. The company declared a handsome dividend for its stock holders as well as generous bonuses for its salaried and hourly workers. Warren-Ehret's profits also hit a record high in 1948.[16]

Although work was available, success like this did not come knocking at a contractor's door. Marketing became a more important part of even the small contractor's business during the 1950s. To insure his share of the pie, the contractor had to advertise. There were many new competitors in the industry and a contractor had to be sure that potential customers could identify his firm. One sales method, perhaps the most direct, was direct mail with telephone follow-ups. URCA President Art Biebel, with the aid of one saleswoman, sold $6,000 of roofing in 1945 using this method, all without leaving the office.[17] Nicholson and Galloway also used direct mail solicitations, directing them to clients that owned types of buildings in which the firm specialized. In the 1950s, for instance, letters were directed to owners of ecclesiastical buildings, universities, banks and other large organizations.[18]

Newspaper advertising continued to be a popular method of making a contractor's existence known, as well as advertising in specialized journals (if one had such a specialty). Because they worked frequently on church buildings, Nicholson and Galloway advertised in *The Official Catholic Directory* of 1940.

Some contractors believed contracts came from a good image. They argued that one of the easiest and most logical, yet very effective, ways to create a good advertising image was to keep all equipment spotlessly clean. This supposedly created a positive impression with all those who saw the equipment around town. A proponent of this approach was F. J. A. Christiansen. Christiansen's dark green trucks traveled around Milwaukee, creating an image of quality for many years. "My grandfather actually started that," Robert Christiansen, F. J. A.'s grandson, reported. "He was very particular about the harnesses for the horses. The collar, the strapping and all leather had to be blackened and shined and the brass polished every week. The wagons had to be washed and polished every week. Through the years, this . . . was taught to my dad and he taught it to me."[19]

Despite the overwhelming success experienced by the construction industry during this period, it was not without its inter-trade rivalries. In 1947, URCA President Cockrell established the National Labor Relations Committee that was to meet with the union and help assure "stabilization and standardization" throughout the industry. It was also to try and work out ways to protect the roofer's turf in disputes over jurisdictions, especially with the ever-aggressive carpenters' union.[20]

Unlike the Depression, where fellow workers understood a trade claiming as much work as possible, roofers in the 1950s reacted sharply to the carpenters' post-war attempts to expand their jurisdiction to roofing work. Carpenters mixed their power play with a sense of history. To the carpenters:

"In taking work away from the roofing industry (and it affects employers as well as employees), the carpenters state that they are merely 'taking back' work, on the theory that since all structures in colonial days were constructed of wood, their jurisdiction continues even though materials change."[21]

The roofing union remained unimpressed and grew increasingly angry over the carpenters' attempts at expansion.

These disputes eventually took an ugly turn. In Gary, Indiana, there was a dispute between carpenters and roofers over who should apply asphalt shingles. Roofers had been awarded the contract, but the carpenters wanted the work to give to unemployed carpenters. The carpenters were so vehement in this instance that they even tore sheathing from some roofs to halt the roofers' work. Not only did this action incur the animosity of homeowners, it was a criminal offense. The dispute eventually was settled by the National Joint Board for the Settlement of Jurisdictional Disputes, but until they had a chance to hear the case, the roofers continued working under the terms of their original contract.[22] Protests, threats and counter-threats continued.

Despite the logical rationale that roofers should apply shingles, these disputes were not always decided in their favor. In 1953, for example, roofers and carpenters were struggling over who would apply strip shingles on a job in Alabama. The National Joint Board awarded this project to the roofers. In Washington, on the other hand, carpenters were awarded the application of asphalt shingles because it was the "established practice in the area."[23] And so it went; the carpenters lost a few disputes, but won enough to encourage them to continue their efforts.

Throughout this period, the construction industry helped build suburban America. At the war's conclusion, many people were able and willing for the first time in 15 years to buy their own houses. Attractive houses in the city were, in many cases, too expensive. Others shied away from congested city neighborhoods and began looking outside the urban areas. The automobile gave people the opportunity to live in the countryside, but still work in town each day.

Quick to follow up on a trend, real estate developers began buying whole tracts of rural land near the nation's urban centers. Subdivisions were built—row after row, street after street of houses for potential homeowners. To minimize variation in construction and thus maximize profits, the developers would fill an entire subdivision with only a few different house models. Suburban development made the developer happy; it made the new homeowners happy; it made the automobile industry happy because most of the new suburban homeowners, forced to leave a car for the wife and children, needed a second car to commute to and from work.

It was the cities themselves that lost the most in this new arrangement. By losing these new homeowners to the subdivisions, they lost a vibrant force for change and renewal within their corporate limits. Cities were becoming a place to visit, not to live. As a result, urban housing stock began to deteriorate. New commercial and retail areas began springing up around the new suburban developments, thus negating the need for people to come to a central business district to shop.

Television was another new development in the late 1940s and early 1950s. Invented in the 1920s, the concept of television was thought to be no more than a novelty; besides, nothing could replace the radio. By the late 1940s, however, the radio networks were willing to gamble and began providing commercial television programming. With shows like the Texaco Star Theatre, starring the irreverent Milton Berle, television quickly won a permanent spot in the family living room. The nation's rapid acceptance of this innovation was unparalleled in American history. In 1950, only 9 percent of the households across the country had television. In five short years, that figure was 64.5 percent and by 1960, it was 87.1 percent.[24]

Television provided mixed blessings for the roofing contractor. To get the best reception, the set had to be connected to an outdoor antenna, and most frequently that was placed on the roof. As a result of improper antenna installation, roofing contractors got many roof repair jobs. But leaks caused by the installation itself, as well as the wear and tear on the roof from antenna installers going back and forth, provided many headaches for the contractor; these jobs were small and bothersome.

One of the most significant changes to the post-war roofing industry was the increased mechanization of application. Technology had done little to change methods of applied built-up roofs for almost 100 years. As late as the 1940s, contractors routinely relied on the "spit test" to determine if the pitch was at the proper temperature for mopping. If the spit hit the pitch crackling and sizzling, the kettle was too hot and the tar was pulled from the fire.[25] Labor shortages during the war sped the acceptance of the oil-fueled tar kettles, which had been introduced in the

1930s. Oil burners were more efficient than wood burners because they could heat the bitumen to the required temperature in 10 to 15 minutes. Wood burners took anywhere from one to three hours. Not only did the oil burners heat the bitumen more quickly, they allowed the crew to work more efficiently:

"The average labor saving alone with oil-burning kettles has been definitely checked at approximately 20%, as the speed of the roofing crew depends on the speed at which the mopper works, he being the 'key' man, and because he can get over so much greater area when the 'hot stuff' is really hot and his mop slides along easily instead of dragging and pulling when the 'stuff' is merely warm or cold, he naturally speeds up the rest of the crew. Remember, oil-burning equipment crowds the men instead of the men crowding the equipment, as is generally the case with wood, coal or coal heaters. No more delays of the entire crew sitting about wasting time and money because of the old excuse, 'waiting for the hot stuff'."[26]

Oil burners also aided the quality of the roof. The application temperature of bitumen is a critical factor in applying a good built-up roof. If it is too cold, the bitumen can become very thick and will not cover very well. Conversely, if it is too hot, it becomes too thin and will bubble up, trapping air between the plies. Oil burners helped assure a high-quality roof, because their temperature is more easily controlled than those of wood burners.

Mechanization increased in the post-war years. The new machines helped do just about everything involved with laying a built-up roof. In 1945, Charles Smith invented a completely self-contained felt layer. It included a fire gun, asphalt kettle, mopping and brush attachments, and was powered by a small gasoline engine. Needing only an operator and one or two people to keep it supplied, the machine could do the work of 15 men, laying 300 single-ply squares in a day.[27] In 1947, the Matt Coil-less Burner Company of Chicago introduced a felt layer that was not motorized. It had a hot-stuff reservoir, a feeder valve regulating the hot-stuff flow, a mop and a roller. Despite the lack of a motor, it was " . . . as easily pulled as a baby buggy." On a slightly improved model in 1948, this felt layer was able to lay a 150-foot strip of felt in 29 seconds as well as 200 squares of five-ply roof in eight hours.[28]

In reroofing jobs, one of the great problems was getting the old roof off and knowing how long that would take. Some roofs came off readily; others were still well-adhered and required considerable work. In 1947, the Benmore Slag Scraper and Sweeper was developed to eliminate that problem. It was capable of anything from sweeping old gravel from a roof if one were to be recoated, to completely tearing off a roof down to the deck.

"With the mechanical scraper, it is possible to accurately predict the labor cost as the machine's rate of progress does not depend upon weather or

Contractors and guests pose during the annual banquet at the 1963 NRCA convention. The trade show featured 70 booths. (In 1987, there were more than 800.) Dr. Wilford White from the Small Business Administration delivered the keynote address on "How You Can Grow and Make More Profit."

upon how tightly the gravel or slag adheres to the roof."[29]

A year later, a scraper was put on the market by G. H. Tennent, Minneapolis, that was equipped with a 16-inch drum and a variety of cutters suited to almost any job. Operating at

1,500 revolutions per minute, the machine was capable of virtually any task required for reroofing preparation. A pitch-and-gravel roof on a wooden deck was among the types of roofs for experimentation. It was less than two years old and virtually unremovable by hand. The Tennent scraper removed it down to the deck at a rate of 810 square feet per hour.[30]

Mechanization was quickly embraced by the roofing contractor as the results of a 1955 survey of 250 contractors by the *National Roofer* shows. Between those questioned, they had 247 kettles, 212 hand hoists, 180 power hoists (for lifting material to the roof), 240 trucks, 124 conveyors (also for getting material on to the roof), 144 felt layers, 121 power scrapers, and 86 roof sprayers.[31]

Building trends during World War II and in the post-war era encouraged contractors to mechanize. Large defense plants gave many contractors their first taste of mechanization during the war. A phenomenon of suburban development was the shopping center, frequently a large building in which many stores were concentrated, thus providing a one-stop shopping area for the new suburbanite. These buildings could have roofs that covered several acres—roofs that, without the aid of mechanization, would have been quite difficult to install. In Minneapolis, for instance, Nees Brothers applied two roofs on warehouses, one being approximately seven acres and the other being about six acres. In Norfolk, Virginia, an eight-acre Barrett roof was applied to a dockside warehouse of the Norfolk and Western Railroad.[32]

The United Roofing Contractors Association underwent some significant changes following World War II. To enhance the industry's reputation, URCA in 1948 inaugurated an annual award program. Each year, URCA would select an individual who had contributed greatly to the welfare of the industry. The award was the Association's effort to improve roofing's image in the building sector. The first award was given to Joseph A. Piper, a North Carolina contractor who helped hold URCA together during the Depression. In later years, the award was named after Piper "in grateful appreciation of his long and outstanding contribution to the roofing industry."[33] Of more immediate significance was the decision to change the Association's name. At the suggestion of James McCawley, executive secretary of URCA, the name was changed to the National Roofing Contractors Association. The name United Roofing Contractors Association, McCawley argued, had made sense in 1921 when the old National Association of Master Slag and Gravel Roofers joined with the Associated Roofers of America. But since that time, URCA had grown to become a truly national organization. A name change, it was felt, would recognize this fact and help give the Association more prestige in the construction industry. The change was adopted at the Dallas convention in 1949.

The newly christened National Roofing Contractors Association had some very rocky early years. In 1951, James McCawley's duties as the fiery editor of the Association's *National Roofer* and as the chief employee of NRCA began to conflict. For his part, McCawley derived great personal satisfaction from the writing, research, and editorial functions of *National Roofer*. He was loathe to turn his back on journalism. He instead advocated increasing his staff so that both jobs could be done properly. The Board baulked at further expenditures and ordered McCawley to discontinue the magazine if it would conflict with his work as executive secretary. Instead he tendered his resignation and made an offer to buy the magazine.

During the height of the Baby Boom, families were moving into tract housing as fast as it was built. The roofer population exploded, reaching 50,000 by 1950.

Both were accepted. After 13 years, NRCA needed a new executive secretary and McCawley had a new journal to publish, the *National Roofer and Building Improvement Contractor*.[34]

The dispute left NRCA weak and without effective leadership. C. N. Nichols, head of the Northeastern Roofing, Siding, and Insulation Contractors Association, sought to use this problem as leverage to create a giant trade association. Nichols approached the NRCA Board and proposed a merger with NERSICA. It was a critical moment in the history of the Association. A union with NERSICA might have produced a stronger organization for roofing contractors, but there were risks as well. NERSICA, with Nichols at the helm, would have been the dominant partner in the merger. That association was already beginning to broaden its focus to the entire building improvement field. Roofing might have lost its distinct national mouthpiece if the proposed NERSICA-NRCA union had gone through.[35]

As it was, the dispute with McCawley cost NRCA a great deal in efficiency and the membership lost much in service. The new executive secretary was Carroll C. Figge, a flamboyant manufacturer and no stranger to the Association. Despite his familiarity with the organization, having been an officer for some years, Figge was not the best Association manager. Association services dropped to the point where the only material for distribution to members were NRCA decals. Figge also put out bulletins, on a more or less regular basis, that expanded on various industry topics. They were not Association positions, however, but his opinions—editorial comments of a sort.[36]

By 1957, the Association was clearly in a bad way, and Calvin Bowman, NRCA president-elect, wanted action. Bowman was an old hand at management who could size up a situation and do what needed to be done. Raised in Oklahoma, Bowman's father, a concrete contractor, moved the family to Florida during the land boom of the 1920s. He went broke in the 1926-27 crash, so Calvin went back to Oklahoma looking for work. He went to Standard Roofing Supply, where his father bought many of his materials, and was hired as a flunky at 25 cents an hour. He ultimately became president of the company.[37]

Bowman felt that Figge had to go to get the Association back on the right track. His acceptance of the Association's presidency was on the condition that Figge tender his resignation. Figge did. The 1957 convention, the Association's 70th, was his last.[38]

After several months of searching for a new executive secretary, Bowman and members Myron Powell and Clyde Scott interviewed Fred Good. The committee was impressed. Good, who was employed by a multiple-management firm in Chicago, began working part-time for the Association in 1957.[39]

The Association was not the only facet of the industry having problems in the 1950s. In March 1958, the Senate Subcommittee on Antitrust and Monopoly and the Committee on the Judiciary began investigating the policy of pricing asphalt roofing products. The objective was to determine if a zone pricing system had been developed by large manufacturers to minimize, if not totally eliminate, competition from new product manufacturers in any given zone. If this was so, the Subcommittee was further questioning whether current anti-trust law was adequate to deal with the problem. Testimony was derived from five sources: officials of the Leopard Roofing Company and the Volasco Products Company, both small companies attempting to start up; and the Lloyd A. Fry Roofing Company, CertainTeed Corporation and the Ruberoid Company, all three with 38 plants between them nationwide.[40]

Weber Pharis, president of the Leopard Roofing Company, testified that, after working as a salesman in the asphalt roofing industry for several years, he was convinced that it would be a profitable small manufacturing area to enter. Taking his savings of $43,000, he embarked on such a venture, building a factory in Meridian, Mississippi to produce saturated felt. The factory began operation in August 1955, with the production of various weights of saturated felts. Pharis had planned to set his prices at a cost-percentage basis. He was going to run a low overhead operation and wanted to pass the savings along to his customers, even if that was at a price below that set by the industry. Lloyd Fry himself indicated that to gain a foothold in the market, "any newcomer would have to sell under the price of the major."[41] Pharis had a successful operation, as demonstrated by his ability to pay off part of his mortgage in the fall of 1956.

Shortly after paying on the mortgage, however, the industry offered a discount on saturated felt, Leopard's only product, in the geographic area he served. Leopard, its sales reduced dramatically, then expanded its product line to include shingles and roll roofing, items that were not discounted. Production on those items was inaugurated in June 1957. A month later, the industry offered additional discounts on saturated felt, as well as on Leopard's new products of shingles and roll roofing in his marketing area of Mississippi, Alabama, Tennessee, Georgia and northwest Florida. To try and lower his costs and compete with the discounts, thus saving his business, Pharis built his own dry felt mill.[42] At the time of his testimony, the mill was under construction.

Lloyd Fry, testifying as the asphalt roof product manufacturing leader, allowed that until February 1956, he had sold his materials at a 5 percent to 7.5 percent discount below major competitors. He was passing along some overhead savings. Yet, in 1958, "Mr. Fry now believes that it is the right of, and proper for any established producer to meet the price of the newcomer . . . and reserves the right to undercut a competitor in a particular section of the country if he sees fit." Fry went on to contend that the asphalt roofing industry was "greatly overexpanded by many opportunists who were looking for quick profits." Mr. Pharis, he continued, was on "unsound economic ground having started [his] new plant" and "actually got started by taking business at lower prices," thus helping create "this chaotic condition."[43] Both Ruberoid and CertainTeed echoed Fry's contention that they had a right to cut prices to maintain their market share.

The Senate Subcommittee concluded that the pricing practices employed by the major manufacturers was typical of that used by early trusts, to force their competitors out of business.[44] While acknowledging that anti-trust laws were not suffi-

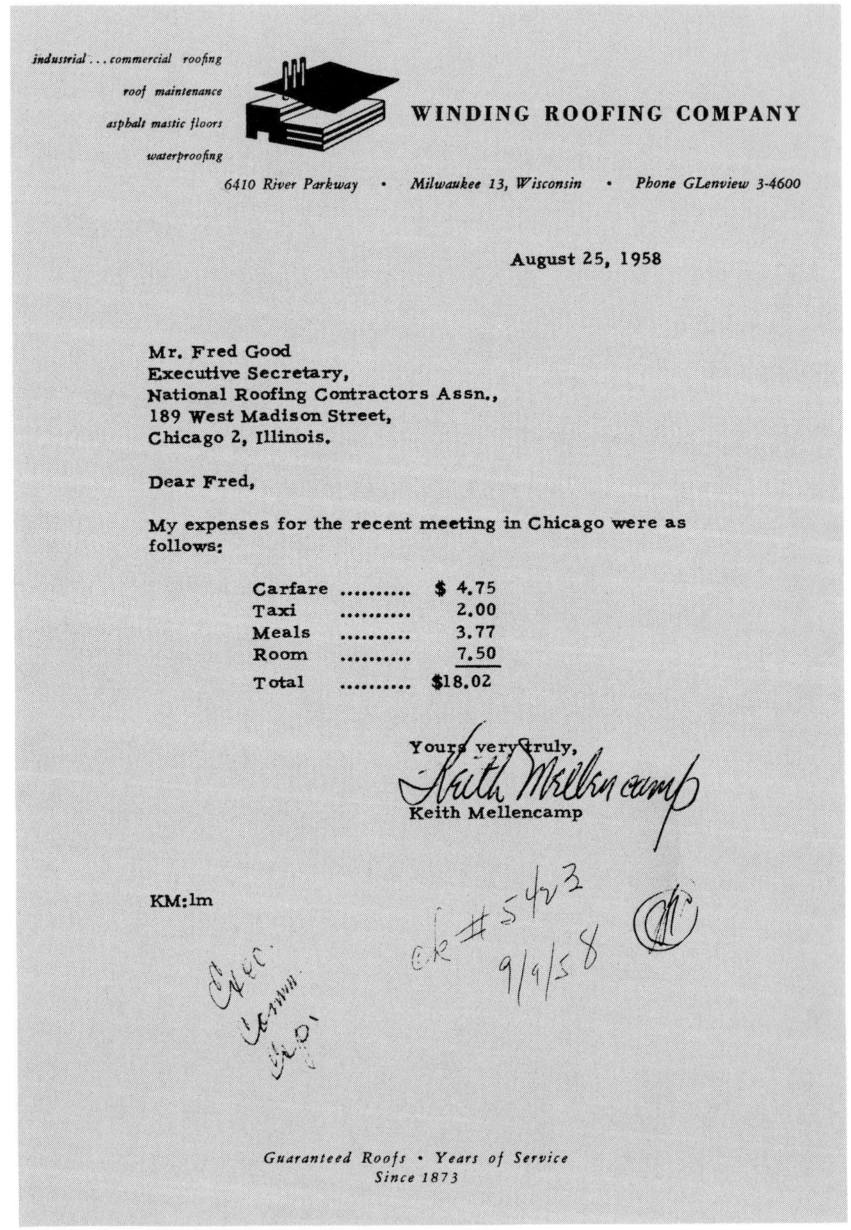

industrial . . . commercial roofing
roof maintenance
asphalt mastic floors
waterproofing

WINDING ROOFING COMPANY

6410 River Parkway • Milwaukee 13, Wisconsin • Phone GLenview 3-4600

August 25, 1958

Mr. Fred Good
Executive Secretary,
National Roofing Contractors Assn.,
189 West Madison Street,
Chicago 2, Illinois.

Dear Fred,

My expenses for the recent meeting in Chicago were as follows:

Carfare	....	$ 4.75
Taxi	....	2.00
Meals	....	3.77
Room	....	7.50
Total	....	$18.02

Yours very truly,

Keith Mellencamp

Keith Mellencamp

KM:lm

ck # 5423
9/9/58

Guaranteed Roofs • Years of Service
Since 1873

Keith Mellencamp (Winding Roofing Company, Milwaukee) submits his expense reimbursement form for his 1958 trek to Chicago: $18.02 for transportation, meals and lodging.

cient to deal effectively with these situations and recommending further study on how to remedy the problem, the Senate hearing did serve to illuminate a rather shady practice in the roofing supply portion of the industry.

This period also witnessed a significant drop in roof quality control as the authority exercised by manufacturers over their franchises was questioned. From the advent of the Barrett Specified Roof and Barrett-Approved Roofers in 1916, a franchise was the key to a contractor's success. Whenever an architect called for a Barrett, Carey or Johns-Manville roof, for instance, a contractor could not bid on the job unless he had the appropriate franchise. It was essentially his license to work. Over the years, manufacturers used the franchise as a method of keeping contractors in line and making sure that their work was meeting the appropriate standards. If a contractor erred, his franchise was removed. He was out of luck, if not out of business.

This practice meant that only a handful of roofing contractors could bid on any one job, to the exclusion of all others. This appeared to be monopolistic, and the government investigated the arrangement and ultimately issued a consent decree. This action virtually ended the manufacturers' control over who could obtain one of their franchises and allowed practically anyone to get one. Because new franchises could be so easily obtained, the threatened loss of one was no longer incentive enough to insure proper quality on the job. As a result, roof quality and the industry's reputation began to drop precipitously.[45]

Except for the mechanization introduced after the war, few innovations occurred in the composition roofing industry in its first 100 years. In 1960, the standard composition roof was still a four-ply roof, but now the industry proposed reducing the number of plies. Initially, the four-ply system was replaced with a base plus three-ply. That soon gave way to a base and an additional ply of coated felt. This was the manifestation of the ill-fated two-ply systems, a system where two plies were supposed to be equal to the original four. The new roof caught on quickly, but one major problem existed: it had not been tested over time. Within two years of application, many of the roofs were leaking and failing. The industry had embraced an untested product, and was now paying the price. Two-ply systems were an unqualified disaster. Manufacturers quickly removed them from their product lines and contractors stopped applying them, but the damage was already done. Enough roofs had failed with the new system that, along with the lack of quality control provided by the old franchise system, and the questionable pricing policy of the large manufacturers, the industry's reputation was badly tarnished.[46]

Despite these problems, individual contractors continued to prosper. Construction expenditures grew from $30 billion in 1947 to about $100 billion in 1967.[47] The number of roofers grew from 50,000 in 1950 to almost 65,000 by the end of the 1960s.[48] Warren-Ehret sales reached the $4 million level in 1962, and J. D. Candler continued to attest to the prosperity of the times by paying generous dividends and bonuses to its stockholders, officers and employees.[49]

This was the industry's situation when Fred Good took the reins at NRCA in 1957. One of Cal Bowman's organizational changes during his tenure as president was the creation of the Executive Committee. Previously, the Board of Directors, about 30 men, conducted all business. Realizing how inefficient that system was, Bowman created a workable body of four vice presidents, a secretary, treasurer, president and past president. This Executive Committee met on a regular basis and performed in a more orderly manner much of the work that the entire Board had done. Thus, as Good joined the organization, Bowman gave him a more streamlined operation, one that could respond to matters more quickly.[50]

Once installed as executive secretary, Good set out to reform the Association. Membership categories were a mess. Despite requiring fixed dues, the Association pretty much accepted anybody as a member, regardless of what they paid. This was changed. A member had to pay his full dues or his membership was terminated.

Good began making regular trips to Washington, D.C. to meet with labor and construction officials to get an idea of what was going on in the industry and to make the concerns of the roofing contractor known. Perhaps most significant among Good's early actions was the material research and testing program the Association set up with the University of Illinois. The Association believed that one of the general industry problems was contractors' ignorance about the behavior of the materials they were applying. If they did understand materials' properties, they would know how to handle certain situations. The testing program, therefore, was not to develop new material, but to test items already on the market under different circumstances. NRCA's initial expenditure for this program was only $7,000, but it was a giant step in acknowledging a commitment to the industry.[51]

Through these efforts, the Association slowly regained a position of influence. By 1965, George Burrus, president during that year, recalls that NRCA had participated in the compilation of the American Institute of Architects manual, an adjunct to the NRCA specification manual. The

Association leaders were also continuing their technical and research work, as well as implementing a technical assistance program, whereby the membership began sharing some of the results of the research with other contractors around the country.[52] These were all measures designed to invigorate the Association and the roofing industry as it prepared to close the decade.

An example of the roof of the not-too-distant future: a model of the San Diego Convention Center, to be completed in 1989.

CHAPTER 9

Lessons of the past: roofs of the future

"I find the great thing in this world is not so much where we stand, as in what direction we are moving." OLIVER WENDELL HOLMES

The late 1960s saw a stronger and more influential NRCA than at any period in its previous history. J. Roy Martin Jr., former president of the Association, referred to the period as a "golden age." The roofing industry was healthy as the expanding economy of the mid to late 1960s fueled an aggressive construction industry.

But it was not just a matter of a good business climate that allowed NRCA to grow in prestige during this era. The 1960s and 1970s saw tremendous expansion in the size and influence of the federal government. All sectors of society felt the effects of this growth, including the roofing industry. Increased government regulation was part of a complex business environment in the contract construction industry. Since the 1930s, the old URCA and later NRCA had aided members in understanding and benefiting from federal housing acts, wage and price controls, and other legislation emanating from Washington. In recent years, however, lobbying, legislative reports, legal advice and industry research by the national Association have made NRCA an increasingly important part of the organizational structure of the roofing industry.

Initially, the Association's political involvement was restricted to simple endorsements of bills that could benefit its members. These were generally federally sponsored initiatives to stimulate construction or home improvement. The Casey Bill in 1967 is a case in point. The measure provided a $750 tax deduction to homeowners who undertook extensive house repairs.[1] NRCA's Executive Committee debated the legislation and sent a formal letter of endorsement to Washington. That same year, the Association increased its level of political involvement when Past President John Reuter testified on behalf of the Leggett Bill before a congressional committee. NRCA's interest in this bill was keen, as it was " . . . to promote fair competition among subcontractors and to prevent bid peddling on public works contracts by requiring persons submitting bids on those contracts to specify certain subcontractors who [would] assist in carrying them out."[2]

Perhaps NCRA's most active and prolonged period of political involvement came with the passage of the Occupational Safety and Health Act (OSHA). Perceiving drastic ramifications for the industry if the Act passed, the Association kept membership abreast of the Act's status throughout Congress' deliberations. OSHA was signed into law on December 29, 1970. President Nixon proclaimed that it was the "culmination of the American System at its best . . . " The labor secretary was equally laudatory, saying, "We plan to launch its administration with all the vigor and momentum we can generate. From this day forward, the health and safety of the American worker becomes a top priority activity."[3]

OSHA was a program designed to ensure that the health and safety of America's work force was not jeopardized by hazardous conditions in the work place. An army of OSHA inspectors began visiting factories, construction sites, retail outlets and offices across the country. Where conditions warranted or where corrections were not made, fines of up to $10,000 were levied. Because roofing by its very nature is dangerous, NRCA anticipated frequent visits to the membership and published a 14-page article in its periodical *Roofing Spec* that discussed the act in great detail.

NRCA was right to be concerned about OSHA's effect on its membership. Many contractors had a casual attitude about job safety. Old-time contractors felt a certain amount of hazard went with the job. At annual meetings, they would often trade stories over beers. Thomas McClain, a Galesburg, Illinois contractor, used to tell about the time he fell off a high roof. He tumbled 20 feet through the air before crashing through the limbs of a tree. Fortunately, he was caught in the network of branches. After collecting his senses, McClain climbed down from the tree and proceeded to go back to work. Charles Orth was also typical of the tough, older generation of roofing contractors. An Ohio contractor, Orth had a long and prosperous career installing mostly residential roofing. He managed his own contracting firm for 55 years. At the age of 71, he still worked on the roof with his men. He might have lost some of his agility by then, for while working on a high-roofed barn, he slipped and fell 18 feet. Gingerly, the old man got up and checked for broken bones. Finding none, he promptly went back to work.[4] But OSHA was unimpressed with this "macho," stiff-upper-lip attitude toward safety.

To ensure across-the-board compliance with

its guidelines, OSHA selected "target industries" to police thoroughly. The agency took on the five worst offenders, those with the highest injury and death rates. Regrettably, the roofing industry fell within this category. NRCA members began to be inspected by OSHA investigators within six months of the Act's implementation. Among the common items checked were chains and pullies, guards for power saws, and proper grounding of electric tools.

In 1971, a roofer was killed at Emory University in Atlanta. He was working on the ground when a trolley track hoist was pulled from a roof and struck him on the head. The resulting OSHA investigation resulted in a $650,000 fine for the roofing contractor for failing "to inspect slamping screws on [the] rear leg of the assembly of [the] trolley track hoist with sufficient frequency to insure [the] hoist was secure."[5] The contractor appealed the fine, but because of incidents like this, OSHA realized that shops were not the only place to inspect. They began visiting construction sites, too.

NRCA accepted OSHA's existence and tried to prepare the membership to deal with it. In 1972, the Association put out a manual telling contractors how to deal with OSHA inspections. One of the suggestions was to arrange a preparatory inspection by an insurance company or loss control expert.[6]

Despite its efforts to live with OSHA, the Association did not relax its lobbying efforts to either eliminate the agency or at the very least make the regulations easier for its members to comply with. Individual contractors also took action. A suit in 1972 by Lance Roofing Company of Atlanta challenged the constitutionality of OSHA by arguing that its fining procedure deprived one of due process.[7] NRCA contributed to Lance's efforts and urged other members to do the same. Congress also considered amendments to the Act from time to time and the Association did not miss a chance to argue on behalf of its members. In 1972, Bruce Martin, NRCA assistant executive manager, testified as an expert witness to both the Subcommittee of the House Select Committee on Small Business and the House Select Committee on Labor concerning OSHA. He presented arguments for modification of the Act based on NRCA member complaints and suggestions. According to Martin's testimony, the basic problem was "that the standards themselves are unnecessarily complex and unnecessarily burdensome and they are applied much more broadly than they were intended to be when they were written."[8]

The relationship between OSHA and NRCA was not always combative. The agency and the

The NRCA logo was refined as the Association came of age.

1960 *1970* *1980*

Association were both attempting to improve the quality and condition of the work place. Although NRCA disagreed with many OSHA actions, the agency stood to benefit from NRCA's long-term commitment to increase professionalism within the industry. On several occasions, OSHA praised NRCA for having done "an impressive job responding to the challenge." But generally such praise came within the context of conflict.[9]

As the 1970s wore on and OSHA became a well-entrenched bureaucracy, NRCA tried to modify regulations that it considered impractical. One such regulation called for "every open-sided floor or platform 6 feet or more above an adjacent floor or ground level shall be guarded by a standard railing or the equivalent . . . on all open sides, except where there is entrance to a ramp, stairway or fixed ladder."[10] In its literal interpretation, this regulation was applied to those working on roofs; thus contractors were being cited for not putting fences around the roofs that their workers were on. This regulation was not only impractical from the contractor' standpoint; it did not take into consideration the hazards threatening those who would be responsible for attaching the guard rail to the sides of the building or the structural requirements for installation.

NRCA evaluated the problem. The Association conceded that men did fall off roofs, and decided that a major problem was simple lack of attention. Workers were not aware that they were close to an edge and as they were working, would sometimes slip over. In response, the Association conceived of the safety line—stanchions with some type of line strung between them, placed about 8 feet from the edge of a roof. It would warn workers that they were approaching the edge and thus prevent them from falling off.

NRCA certainly did not want to go into the business of writing OSHA regulations that the membership would then have to obey. But in the safety rail case, the Association was faced with a totally impractical requirement; it needed to try and make the regulation as workable as possible for both the contractor and OSHA. Largely because of the efforts of Philadelphia contractor Robert Linck, who lobbied OSHA to accept the warning line concept as the safety standard for flat-roof edge protection, it became an enforceable regulation in November 1979.[11]

NRCA expanded its commitment to political activism in 1976 when it hired a representative in Washington, D.C. He was William S. Bergman, president of William S. Bergman Associates, a government relations firm. Bergman was to establish personal contact with legislators and discuss items that were of direct interest to the roofing industry. The Association was assuming a greater role in influencing the laws emanating from Washington. In addition to hiring Bergman, it urged its members, in light of the 1976 elections, to write their representatives about the minimum wage law, Workmen's Compensation, energy, regulatory reform, consumer protection, unemployment, public works jobs and (as always) OSHA.[12]

The energy crisis of the 1970s affected all sectors of the economy. But for the roofing industry, the high price and scarcity of oil had a tremendous impact on the cost of the components on which contractors relied. Shingles, saturated felt, and asphalt were all derived from petroleum. In a matter of months, material shortages arose where once there was abundance; where material was available, prices had quadrupled. This was particularly hard on the contractors who had signed fixed-rate agreements because the price increases could not be passed on to the consumer. James King, NRCA president when the shortages began, took steps to secure what information about pricing and material availability he could and passed it along to the membership. Hill's successor, William Steinmetz, set up a formal structure to study the oil crisis and determine industry priorities. These studies and the continued monitoring of OSHA strained the Association's ability to meet its expenses. NRCA was carrying deficits larger than any previous cost overruns.[13]

Slowly, the Association began to piece things

together. An Energy Committee was set up to continue to get information to contractors about the prices and availability of materials. Also, as different grades of crude oil were imported from other countries to alleviate the shortage, NRCA and the Asphalt Roofing Manufacturers Association (ARMA) began a joint program to evaluate the qualities of each crude; the objective was to discover the differences and to compensate for them to produce a uniform grade asphalt for the industry.[14]

The oil shortage created an energy awareness across the country. Not only were people quickly changing to the smaller, more fuel-efficient cars, they were also turning thermostats down, turning lights off and insulating their buildings. One of the building components that needs insulation is the roof; so roofing contractors took an active role in promoting energy conservation. On July 10, 1975, NRCA representatives met with White House officials about the industry's role in the nation's energy program. Donald Ogilvie of the Office of Management and Budget and Roger Sant of the Federal Energy Administration were present. The meeting reflected the gloomy prospects of the industry. Sant discussed oil reserves, which were estimated to last only 25 to 40 years. He said that to meet future demand, alternate energy sources must be used. He predicted that oil prices would keep rising. On the positive side, he also outlined planned tax incentives for energy-efficient buildings.[15]

NRCA and the roofing industry responded to the Administration's challenge by creating a public relations campaign: "Good Roofs Save Energy." They began trying to sell energy-efficient roofs. Promotional literature was prepared for members to send to potential clients. An *NRCA Energy Manual* was drafted for the membership, with preliminary copies distributed at the 1976 convention in Phoenix. The *Manual* advised contractors on such items as calculating savings potential based on insulation used, in energy units saved as well as dollars. During the 1977 convention in Atlanta, the energy program dominated discussion. NRCA members participated in three radio talk shows and a television show to discuss roofs, insulation and energy efficiency.[16]

President James King talked about the potential for innovative roofing systems. "I suppose someday," King said, "another material must replace asphalt and coal tar pitch as common waterproofing materials . . . It would not seem possible today or in the immediate future, however, that a new material competitive in cost and equal or better in quality is forthcoming."[17] The oil shortage began the search for alternatives to asphalt in roofing products, for even when the shortage ended, the industry knew that the petroleum sources were limited.

It is often argued that the immediate need for an item creates the technology and the will to produce it. Applying this to the lack of change in the built-up roofing industry in the 125 years preceding this period, it can be argued that the industry continuously offered a product that worked; technology was not forced to change it. In the mid 1970s, however, the roofing industry directly challenged technology to produce new products—materials that were no longer as dependent on asphalt and therefore on oil as the previous ones.

A major result of this technological challenge was the development of the single-ply roofing membrane. Although some of these are asphalt-based, such as the modified bitumen roofs, others are constructed totally of synthetic materials.

One of the earliest single-ply roofs dates back to 1957 and the Ingalls Hockey Arena on the Yale University campus at New Haven, Connecticut. The arena was designed by architect Eero Saarinen, a Finnish immigrant born in 1910 who came to this country in 1923 with his family. Saarinen would play a key role in the development of the single-ply systems. His father was Eliel Saarinen, a well-known architect who, even though he placed second in the competition to design the Chicago Tribune buiding in Chicago, had a more lasting effect on architecture than did the winner of the competition. The Saarinen family settled in Michigan and Eero obtained his architectural education at Yale from 1930 to 1934.[18]

Eero Saarinen was an innovator. His architecture "was known for its craftsman-like beauty as well as its utility and symbolic function." It was "a heroic, monumental architecture but one freed of the literary and archaeological conceits of the previous two centuries."[19] His scale was massive and it was new. Among Saarinen's most widely known works are the TWA terminal at Kennedy International Airport, New York; Dulles International Airport outside of Washington, D.C.; and the St. Louis Arch.[20]

Saarinen felt that a building's design should convey a sense of what the building was about, what function it was to serve. For the Ingalls Arena he " . . . sought to express . . . the excitement of ice hockey . . . creating a vessel for sport . . . "

"The Yale rink is spanned lengthwise by a center parabolic supporting arch in reinforced concrete; tension cables are suspended from the arch and are anchored laterally in low concrete walls, which enclose the building and are of similar curvature on plan. At each end the arched spine is extended outward in a reverse curve to form a cantilevered beam supporting entrance canopies. Once a student of sculpture, Saarinen contrasts curves with countercurves, concave with convex."[21]

The roof of the arena consisted of large timbers that were hung on the cables. It was not a rigid roof; it seemed to float in the air. Saarinen had looked for a material to cover the roof but had virtually no luck. He had heard about, and went to E. I. Du Pont to look at, a rubber skin made of neoprene that could be applied to the roof deck. During his deliberations about the roof, Stanley Warshaw and Blair Lamont, partners in the firm of Technical Coatings, Inc., visited the construction

Eero Saarinen designed the first building to use a single-ply roof: the Ingalls Ice Arena at Yale University, affectionately known as "The Yale Whale."

site. They proposed to cover the deck of the roof with neoprene, just as Saarinen had been contemplating. The rubber would be elastic enough, they reasoned, to tolerate the swaying of the roof, yet retain its watertight characteristics. Without first sanding down the entire wood roof deck, however, they found the proposed method of direct adhesion to be impractical.[22]

The neoprene would work; the problem was attaching it to the roof. Warshaw and Lamont decided to try and adapt a concept for a roof expansion joint that they had developed to answer Saarinen's needs. Substituting 48-inch-wide rolls of neoprene for the narrow strips used in expansion joints, they proposed to adhere metal strips to the edges of the neoprene that could then be nailed into the roof deck. The neoprene adhesive would then be needed only for covering the strips of metal. Saarinen accepted the proposal but requested that the joints be disguised as a metal roof standing seam.[23]

By the time the roof was complete, more than 15,000 linear feet of aluminum strips had been adhered to the neoprene. The roof held up reasonably well, but was susceptible to ozone-induced deterioration. It also cracked at places of high stress, thus necessitating patching jobs. The membrane itself was floating, attached to the deck only where the strips were nailed down. An unforeseen problem with this was condensation between the deck and the membrane as a result of the humidity of the hockey arena.[24]

After Saarinen's pioneering work, the use of single-ply roofs increased dramatically. In 1977, only 4 percent of the nation's roofs were covered with single-ply; by 1980 it was 20 percent. A sign of its growing popularity was the prominent role of single-ply systems at the 1979 NRCA convention in Houston. The Association sponsored a session on single-ply to familiarize contractors with a system that could command a major market share in the future.[25]

Single-ply roof coverings were not the only result of the need to find alternatives to asphalt-based products. There was also an increase in the use of metal roof coverings.

"Materials that have been in use for years are making a comeback," Sidney Epstein of Follansbee Steel says. "Terne, terne-coated stainless steel and copper have withstood the test of time. The industry is refining old products to meet modern requirements."[26] While many metals are regaining popularity, most are not in the same form as originally introduced in the 19th century. Frequently, they are alloys or at least the products of new chemical processes that increase their durability.

Terne has been revived, using a base of stainless steel. Copper, of course, is also still being used. But instead of the 16-ounce material used for more than 100 years, the new material is lightweight, only 12 ounces, and has exceptionally high strength capabilities. One firm has introduced a complete aluminum roof covering package. Always difficult to solder, the aluminum in this package is fastened with a snap-on batten system.[27]

Zinc, with its extreme cycles of popularity in the 19th century, also made a comeback in the mid 1970s. Zilloy, produced by New Jersey Zinc, was a rather unique reincarnation. The product actually was a zinc alloy and included zinc, copper and titanium. Other metal roofs made appearances by product names rather than metal names. Colorklad, for example, was being produced by Vincent Brass and Aluminum Company in the mid 1970s. It was 24-gauge galvanized steel covered with a resin-base paint unaffected by saltwater, chemicals and other industrial pollutants.[28]

The introduction of portable power-seaming equipment was one of the key factors that made metal roofing in the '70s competitive with built-up roofs. Moving along at 50 feet per minute, power seamers did all the routine seaming, work that once took the major portion of the sheet metal workers' time. On a 1,400-square-foot standing seam roofing job in 1975, the power seamer

Workmen tap down the aluminum strips attached to the edges of the neoprene sheets used on the Ingalls Hockey Arena roof.

completed 2,500 to 3,000 linear feet of seams in a seven-hour day, work that would have taken one man more than a year to do.[29]

The increasing technological complexity of the industry demanded a new professionalism from contractors. The fiasco of the two-ply systems of the mid 1960s had taught many contractors a valuable lesson. Good business practice demanded that they work much more closely with manufacturers and customers to resolve roofing problems. The litigation resulting from the failure of two-ply systems forced contractors who had previously only thought of application to act more like modern businessmen, utilizing marketing and industrial research to attract and reassure customers.

The seed leading to this growth in professionalism was planted after World War II. Armed with the GI Bill, many veterans went to college and earned degrees. Working on roofing crews during the summer, they turned to the roofing contracting industry for full-time employment upon graduation. The result was that many roofing contractors in the 1950s, 1960s and beyond were very well educated. With degrees in business, engineering and other fields, they understood the need for a sophisticated approach to their work.

For the Association, this growth in professionalism directly affected its relationship with its members. Membership had been growing slowly as NRCA re-established itself as a force within the industry, but this growth was tied to what contractors perceived the benefits of joining to be. Service became the operative word for the Association.

NRCA gradually expanded its technical services program. An example of the orientation of these programs was the "Equal Viscous Temperature Concept" programs. Old-time roofers relied on their experience and the famous "spit test" to check the proper temperature of composition roofing adhesives. It was uncanny how a veteran roofer could properly estimate the proper temperature by this homespun method. Unfortunately, not all crews had the benefit of truly experienced men. In order to insure that roofs were applied in a uniformly top-quality manner, contractors needed to be able to insure that the pitch was applied at its "equal viscous temperature." If the contractor could determine scientifically what that temperature was, with the aid of a thermometer, he would be able to consistently apply the adhesive. NRCA sent survey teams on the job with roofing crews. They recorded the temperature at which pitch could be best applied. These results became the basis for specifications which today guide the application of various composition roofing products.[30]

The Association's most important technical aid to the industry was its manual of roofing practice. Begun in 1966, the manual was conceived of as a potential industry-wide standard source. The manual focused on the standard four-ply system. Smarting from the two-ply failure, the Association wanted to insure successful application of the tested four-ply roof. The manual took several years to complete and represented the combined efforts of some of the best roofing contractors in the country. When NRCA published the first working copies of *The NRCA Roofing and Waterproofing Manual* in 1969, it represented the first complete guide to built-up roofing in the industry's history.

During the 1970s, NRCA placed an increasing emphasis on education. In 1978, under President Melvin Kruger's guidance, the Roofing Industry Educational Institute was established "to advance the professionalism of the entire roofing industry through education."[31] It was open to architects, general contractors, roofing contractors, manufacturers and anyone else who needed proper training in roofing practices. The Institute was to be operated as an independent school. When $300,000 of the $400,000 start-up costs were donated by January 1980, a full year's program of 10 seminars was planned.[32]

In 1980, a project that started out as a simple revision of the 1970 manual concluded with another, completely new manual. With more than 500 pages and 200 illustrations, the new manual incorporated the built-up roof manual, waterproofing manual, steep roofing manual, handbook of accepted roofing knowledge, construction details, and NRCA bulletins as well as a glossary and appendix.

The Association, believing that education was the best way to take advantage of the technology available to the contractor, made a new commitment in 1982. Under the guidance of President John Bradford, an Education Department and the Academy of Roofing Contractors was established.

In addition to the safety inspections by OSHA, the oil shortage, the introduction of many new materials and the continuous attempts at membership education, NRCA and the roofing contractors also had to worry about other health and environmental problems during the 1970s and '80s. Air quality was becoming a national concern, as witnessed by the passage of the Air Quality Act of 1967 and the Clean Air Amendment of 1970. There was concern that the emissions from asphalt kettles might be violating those laws, so NRCA created a committee to evaluate the roofing industry's contribution to air pollution.

Asphalt emissions were not only potentially harmful to the atmosphere, certain types were harmful to the men working with them. At the 18th Triennial Convention of the International Union in 1970, Dr. Irving Selikoff addressed the hazards of pitch. Coal tar pitch contains benzopyrene, a substance linked to skin cancer among chimney sweeps as long ago as the 18th century. It is a direct product of burning coal and was deposited on the inside of a chimney. Selikoff conducted several studies of the effects of exposure to benzopyrene, but the findings were inconclusive and no major changes were recommended for roofing practice. Nonetheless, some roofers and contractors became concerned about exposure to coal tar pitch fumes. Fortunately, coal tar pitch is not used exclusively anymore, so that the hazards of roofing are not as severe as they once were.[33]

Another problem that Dr. Selikoff discussed at the labor convention was that of asbestos. " . . . I do not think [it] is as yet very important," he was quoted as saying.[34] Asbestos is only a threat to health if the fibers are airborne and susceptible to inhalation. That is not the case when roofs are applied. Selikoff warned of the possibility of danger when roofs were torn off and dried felt cracked.

Asbestos, hailed as the miracle fiber for the roofing industry when the H. W. Johns Company introduced it in the 1870s, had been good to Johns-Manville, the Johns Company successor. Starting with roofing materials, Johns developed many different products containing asbestos. Among those used in the construction industry were fireproofing and insulation. But the wonder material became a major industry problem in the 1980s.

Johns-Manville entered the 1970s a healthy company. In the first two quarters of 1978, it showed a net profit of almost $60 million. That same year, three federal agencies released reports that predicted that within 35 years, up to 18 percent of the nation's cancer deaths would be related in some fashion to asbestos.[35] Later in 1978, a class action suit for $1 billion was filed by 5,000 shipyard workers at Todd Pacific Shipyard in San Pedro, California, and at the naval shipyard at Long Beach, against 15 firms that used asbestos. The shipworkers claimed that the defendants hid the dangers of working with asbestos from their employees and "did not take those measures necessary for the protection of the health and lives of [the] plaintiffs." Among the 15 firms named were Johns-Manville, Celotex, Philip Carey and Ruberoid.[36]

The Environmental Defense Fund made a very damaging claim against asbestos in 1978. The Fund stated that asbestos sprayed into school buildings for soundproofing, fireproofing and insulation between 1940 and 1973 may have exposed millions of children to the material. They further identified Manville as a leading manufacturer and supplier of asbestos. By the end of 1978, there were 1,000 asbestos cases pending nationwide.[37]

In 1979, a congressional subcommittee voted to establish a $30 million fund to help schools detect and remove asbestos. They planned to recover the money from asbestos manufacturers, primarily Manville. Despite the congressional action, Manville President McKinney told his stockholders that the asbestos-related claims against the firm will be "'far fewer than what has been reported to you." When queried about the firm's financial ability to withstand asbestos claim losses, Vice-Chairman Francis May said that "the general business media would have you believe that our financial situation is such that we are going to have a hard time making ends meet down the road. This is ridiculous."[38] Manville was predicting 1979 earnings to top the 1978 record of $121.6 million.[39]

Things began getting tough for Manville in 1980. A potentially precedent-setting case awarded $1.2 million to a 40-year-old shipworker with asbestosis, a scarring of the lungs caused by asbestos fibers. The defendants argued that the lung scarring was the effect of 25 years of heavy smoking aggravated by a bout with pneumonia, but to no avail.[40]

With the potential for an ever-increasing number of lawsuits, McKinney went on the offensive. "This is the modern version of ambulance chasing. The biggest scandal in our legal system is contingency fees, and that's what all these cases are about." McKinney continued to vent his company's frustration when he conceded that Manville's Dr. Paul Kotin said, "Sure, there can be danger, like everything in the environment, but if it is used appropriately, so that there is low airborne fiber content, there is no danger." But, McKinney said, "Nobody quotes Paul...Instead, they'll go out and quote some pipsqueak from the government."[41] The company continued producing and selling asbestos materials. They vowed all safety methods possible were being used in the manufacture of these products to protect their workers as well as consumers. Manville would "defend the responsible use of asbestos and asbestos-containing products." By the end of 1980, there had been 3,000 asbestos-related lawsuits against Manville and other asbestos manufacturers. Six hundred of those cases had been resolved, with an average award of $30,000 per case.[42]

Things went from bad to worse for Manville in 1981. Again, despite company claims that his lung cancer and subsequent death was the result of heavy smoking, a jury awarded an asbestos worker's widow $850,000, $500,000 in punitive damages. In desperation, Manville appealed the award. Later that year, it had to pay 80 percent of a $410,000 award to another asbestos worker. By March 1984, the company was cited as a defendant in 5,800 lawsuits across the country. By October, it was 8,000 suits and they were increasing at a rate of 400 a month.[43]

To stop the deluge, Johns-Manville had to take a step that was unthinkable only two years before; the company declared bankruptcy in 1982. By filing under Chapter 11 of the Federal Bankruptcy

Code, it asked the court to protect it from further lawsuits and its creditors, and to give it time to work out all its problems. Despite the bankruptcy, Manville assured the roofing industry that the firm was on solid ground and would be there to meet continued demands. Throughout this period, there was a dramatic drop in the use of asbestos. From 1973, use totalling 795,000 metric tons fell to 350,000 metric tons in 1982.[44]

The 1970s also saw an innovative approach to the roofing bond product. It was generally conceded throughout the industry that the bond, as conceived by Barrett in 1916, had outlived its usefulness. Some of the earliest talk of reforming the bond came at the 1948 NRCA convention in Dallas. There Frank McGinley, president of the American Asphalt Roof Corporation, called for the elimination of bonds and the issuance of guarantees limited to two years. Bonds were outdated, he said. They were introduced at a time of poor quality; they were a confidence-builder. But the built-up roof was then more than 100 years old and production and application techniques were advanced to the point where bonds were no longer needed.[45]

McKinley's proposal was a controversial one, but it continued to receive attention throughout the next two decades. The major change finally occurred when Philip Carey, GAF (formerly Ruberoid) and Celotex introduced a new guarantee program in 1970. The guarantee was for 10 years, with an option for renewal for another 10 years. The new programs offered inspection service at specified intervals, thereby encouraging a roof maintenance program. The companies also provided consulting services for designer/architects, an inspection service during application, and obligated manufacturers to investigate problems in a timely fashion. The guarantees were written free of "legalese," and were more specific about what they covered than the old bonds.[46]

As demonstrated by the struggles with OSHA and the energy crisis, the period from 1968 to 1984 was not an easy one in which to do business. To survive the inflation and recessions that occurred in the period, roofing contractors had to make some very deliberate plans to insure their survival. As the J. D. Candler Roofing Company entered the 1970s, President DeForest Candler made a decision that was meant to help insure the firm's future. With all the new roofing companies that had started up since the war, bidding new construction was becoming extremely competitive, so he decided to withdraw from that aspect of the business and deal exclusively with reroofing and roof maintenance. Roof repair was also the strategy that San Francisco's Lawson Roofing Company followed. Frank Lawson remembers his uncle always harping, "Repair, repair, repair. Godammit, repair. Don't put a new roof on." But there were profits in reroofing.[47]

After weathering the economic hardships of the early and middle part of the decade, profits rebounded in the late 1970s. In 1979, Warren-Ehret-Linck (in 1965, Warren-Ehret merged with J. Edward Linck Sheet Metal Works, Inc.) grossed $16 million and showed a net profit of 11 percent, or $1.75 million. In 1979, Candler experienced the highest sales/profit totals ever and authorized $4,500 in Christmas gifts for the year, as well as more than $80,000 in bonuses.[48]

Things changed rapidly, however, and in 1983, after a lengthy and deep recession, Warren-Ehret-Linck filed for protection under the bankruptcy laws. Though not quite as badly off as Warren-Ehret-Linck, Candler also experienced hard times. In addition to the difficult economic picture, the financial strain placed on the company by meeting union wage and work requirements almost closed it down. But with DeForest Candler's decision to abandon the new roofing market, the firm had some leverage for bucking the union. Even in a stronghold like Detroit, union shop agreements basically restricted the large number of firms that performed new construction. Under Candler's new president, Shirley Valade, the union was gradually pushed out of the business. The resulting lower wage rates helped keep Candler profitable.

The late 1960s and 1970s saw a revival of interest in an "open shop" among contractors in all sectors of the construction industry. In 1971, the AFL-CIO reported 24,000 union roofers belonging to 209 locals across the country. Most of these members worked in the Midwest or Northeast and were engaged in large-scale new construction projects. The union had never been strong among small-scale residential roofing contractors, except perhaps for the Chicago and New York metropolitan areas. Jurisdictional disputes, higher wage rates, and strikes all served to sour many contractors on operating a union shop. Organizations such as the Associated Builders and Contractors have been active in advocating open-shop construction.[49]

The unions responded to this trend aggressively. The nature of the response varied from local to local. In some cases, contracts were adjusted and through negotiations, "project agreements" were secured. In other locals, the tools of intimidation were used—boycotts, picketing, even physical violence. In May 1973, the Associated Builders and Contractors charged 17 building trades' unions with conspiracy to violently restrain open-shop contractors. The National Labor Relations Board issued complaints against the international roofers' union.

The worst abuses in the roofing industry came in the Philadelphia area. During the summer of 1972, open-shop general contractor J. Leon Altemose began work on a convention hotel and office complex near Valley Forge. Altemose had long been embroiled in disputes with Philadelphia's Building and Construction Trades Council. He refused to sign subcontractor's agreements that would have limited him to firms employing only union workers. The Trades Council retaliated by attacking the Altemose construction site. Local 30 of the roofers' union arranged for buses to bring 1,000 men to the Valley Forge Plaza site. They spilled out of the buses and destroyed construction vehicles and equipment, eventually causing $350,000 in damage. Later that summer, irate workers, allegedly from the roofers' union, waylaid Altemose outside a downtown Philadelphia bank. In broad daylight, he was beaten. As the project went on, industrial sabotage was rampant at the site. Open-shop employees were intimidated by unionists brandishing baseball bats emblazoned with the Local 30 name. Injunctions and convictions did little to stem the harassment of independent contractors. The actions of Local

The roofing contractor's forebears take a puzzled look at the Illinois Center in Chicago, designed by Helmut Jahn and completed in 1984. An increasingly prevalent question in the construction of today's buildings is, which part of the structure is the roof?

30 seem to have had the tacit support of the Philadelphia mayor's office, which may to some degree explain the roofers' boldness.[50]

Such extreme violence was not the rule in contractor-union relations during the 1970s. But the Altemose case does underscore the depth of emotions roused by the open-shop movement. The roofers' union had been successful in raising the base pay of the average journeyman roofer to a level comparable to that of other construction specialists such as bricklayers, carpenters, sometimes even electricians; this was in direct contrast to the 1920s and 1930s, when roofers were paid considerably less than their co-workers.[51]

The National Roofing Contractors Association tried to steer clear of labor disputes while still serving the contractor's interests. NRCA increased its role as an information clearinghouse and as a voice for professionalization. It began working directly on programs that benefited the roofing contractor: education, material testing, and the development of the litigation center, designed to supply legal assistance to contractors who needed advice.

The 1970s and 1980s have been a period of tremendous growth for NRCA. By 1983, membership had jumped to 2,700 members. Throughout the 1950s and 1960s, the Association's membership had fluctuated around the 700 or 800 mark. Yet between 1972 and 1982, its ranks swelled more than 150 percent.

Such dramatic growth has represented a tremendous challenge to the Association over the past 10 years; its functions have expanded and its staff has grown. Fred Good, initially the Association's executive secretary, later executive vice president, represented a thread of consistency between the modest organization of the 1950s and the dynamic trade association NRCA became in the 1970s. When Good joined NRCA in 1957, he was assisted by two other employees, one of whom was part time. By 1985, the NRCA staff had grown to more than 30 employees, including professional communications, education, meeting planning, technical and data processing personnel. The growth of these departments allowed the Association to expand its services. The annual convention was gradually upgraded. As the program and range of advertisers grew, so did attendance. During the 1950s, if more than 500 people attended, that was considered good; recent conventions have brought together more than 7,000 contractors and manufacturers. Under Good's management, the trade journal *Roofing Spec* became an important revenue-raising venture that allowed other services to grow.[52]

In addition to Fred Good's consistent leadership, NRCA's stable but remarkable growth can also be attributed to its organizational structure. Leadership in the Association is a team effort. The Executive Committee, which is composed of all officers, provides NRCA with its program direction. A member may serve as vice president, senior vice president, president and then finally, immediate past president. Although such a structure makes it difficult for one president to effect change, it does allow NRCA to benefit from the experience of a dedicated elite of contractors.

After Fred Good, William Good may have perhaps had the most impact on shaping NRCA today. A University of Virginia graduate with experience in roofing application, Good joined the NRCA staff in 1973. Initially he, like most staff members of that era, acted as a jack-of-all-trades. He later took responsibility for the Association's quarterly magazine, *Roofing Spec*. Under his direction, the magazine evolved into a monthly, full-color, high-quality publication. Good developed the Communications Department, the Government Relations Department, and the Education Department as he perceived the need for these and the potential services that could result. Good's role grew as the Association grew. By 1978, he had been appointed the Association's general manager, overseeing virtually all of NRCA's non-technical activity. In the early 1980s, he was appointed executive director. William Good's involvement with NRCA continued to 1985, when he left to pursue the establishment of his own business.

In many ways, the current status of the roofing industry is analogous to the situation 100 years ago. Both then and now, there are many competing roofing systems on the market. The ranks of contractors are growing with new entrants into the field, some of whom are new to the construction industry. The labor climate of the 1980s, while not as tense as the late 19th century, is in a state of flux. This atmosphere of uncertainty helped provide the initial impetus for the first roofing contractors' association in 1886, and it seems to have been responsible for NRCA's growth spurt of the 1980s.

In spite of the similarities between the status of contemporary roofing and the industry 100 years ago, one must be careful when drawing on the lessons of the past. Today there is a complexity to business life and a sophistication to roofing contracting that was beyond the comprehension of those pioneer roofers who founded the Gravel Roofers Protective Association. The modern marketplace demands proficiency in all aspects of contracting. The time has long since passed when roofing supplies were the discarded refuse of other industries. Today they are expensive, meticulously specified materials designed to insure not only waterproofing, but fire protection and energy efficiency. The contractor is faced with what might seem to be a bewildering array of product types from which to choose. Marketing schemes, management plans, and industrial relations programs have become vital to survival. NRCA and local associations have found their professional development programs in high demand. Across the United States, managers of roofing firms have sought new approaches to meet the sophisticated climate of contemporary contracting.

But as unique as the current contracting environment is, there remains a remarkable continuity within the roofing business. This characteristic allows the modern contractor to draw upon a rich legacy of experience that can caution as well as encourage.

History's lessons remind the contractor that it took composition roofing a long time to prove itself in the marketplace. The experience of roofers with the infamous two-ply systems during the 1960s underscores the basic fact that whatever system is applied, it is the contractor the client will hold accountable.

The roofing industry remains today an avenue of opportunity for the entrepreneur of modest means. Roofing contracting has always been an inexpensive business to enter. Historically, roofing has been a field well populated with enterprising individuals. But the record of the industry's past 100 years also indicates that survival in the face of business cycle downturns represents a major challenge. The volatile nature of the construction sector insures that the ranks of contractors are periodically purged of the weak or the inefficient. The core of the industry, however, remains solid. Aggressive contractors, responsive to the public's continuing need for quality roofing, have been able to endure the worst business climates—like the great Depression, or the recent cycle of inflation and recession.

Among the most important threads of the past that make up the fabric of roofing today is the industry's strong family orientation. More than 75 percent of roofing contractors today intend to pass their businesses on to family or their employees. Just as pioneer contractor Michael Ehret was introduced to roofing by his father, and the Warrens and Barretts introduced family members to their businesses, family ties help sustain roofing businesses today.

If there is a lesson to be learned from the past, it is that the challenges that face each generation of contractors, as they work to develop their companies, are directly related to the problems of roofing as a whole. Contractors who meet the public's need for professional roofing application, fair pricing, and ethical business practice lay the foundation for their own prosperity and for a positive image of the industry. This is not a concern that can be addressed at one point in history, then neglected or taken for granted.

Professional responsibility demands that each generation strive to leave a legacy of knowledge, craftsmanship and integrity that will challenge generations to come.

And so they have.

CHAPTER ONE: THE SEARCH FOR SOLUTIONS

[1] Andrews, Wayne. *Architecture, Ambition and Americans*. New York: The Free Press, 1978, p.2.

[2] *Ibid*.

[3] Briggs, M. S. *Homes of the Pilgrim Fathers*. New York: Oxford University Press, 1933, p. 122.

[4] Fitch, James M. *American Building*. New York: Schocken Books, 1966, p. 15.

[5] Waite, Diana S., ed. *Architectural Elements: The Technological Revolution*. Princeton: The Pyne Press, 1972, p. 147.

[6] Whiffen, Marcus. *The Eighteenth-Century Houses of Williamsburg*. Williamsburg: Colonial Williamsburg, Inc., 1960, p. 69–70.

[7] Sloane, Eric. *Reverence for Wood*. New York: Ballantine, 1965, p. 90.

[8] *Ibid.*, pp. 15–16.

[9] Glassie, Henry. *The Pattern of Material Folk-Culture of the Eastern United States*. Philadelphia: University of Pennsylvania Press, 1968, pp. 221–223.

[10] Whiffen, *The Eighteenth-Century Houses of Williamsburg,* pp. 69, 202.

Williamson, Scott Graham. *The American Craftsman*. New York: Crown Publishers, 1940, p. 18.

[11] Whiffen, *The Eighteenth-Century Houses,* p. 5.

[12] Whiffen, Marcus. *The Public Buildings of Williamsburg: Colonial Capital of Virginia*. Williamsburg: Colonial Williamsburg, Inc., 1958, p. 227.

[13] *Ibid.*, pp. 57–58.

[14] Beverly, Robert. *The History and Present State of Virginia*. Chapel Hill: University of North Carolina Press, 1947, p. 290.

[15] Whiffen, *The Eighteenth-Century Houses,* pp. 5, 202.

[16] Peterson, Charles E., ed. *Building Early America: Contributions Toward the History of a Great Industry*. Radner, Pa: Chilton Book Company, 1976, p. 137.

[17] Winston-Salem, North Carolina, Museum of Early Southern Decorative Arts. Index of Early Southern Artists and Artisans. *Maryland Herald and Eastern Shore Intelligencer,* 16 November 1802.

[18] *Ibid*.

[19] Peterson, *Building in Early America,* p. 138.

[20] Whiffen, *The Eighteenth-Century Houses,* pp. 5, 68.

[21] Oliver, John W. *History of American Technology*. New York: The Ronald Press Company, 1956, p. 27.

[22] *Ibid.*, pp. 27, 36.

[23] Williamsburg, Virginia, Collection of the Architectural Research Library, Colonial Williamsburg, Inc.

[24] Whiffen, *The Eighteenth-Century Houses,* p. 204.

[25] Williamsburg, Research Library. "British Chemical Manufacturing," 1844.

[26] *Ibid*. "Tarring of Roofs."

[27] Whiffen, *Public Buildings,* p. 75.

[28] *Ibid.*, pp. 45, 212.

[29] Winston-Salem, Southern Decorative Arts. Index to Early Artists. *Raleigh Star,* 28 June 1811.

[30] *Ibid.*, *Raleigh Star,* 8 January 1813.

[31] *Ibid.*, *American Beacon and Norfolk and Portsmouth Daily Advertiser,* 30 November 1819.

[32] *Ibid.*, *Clarian and Tennessee State Gazette,* 29 June 1819.

[33] *Ibid.*, *North Carolina Minerva and Fayetteville Advertiser,* 3 December 1796.

[34] McKee, Harley J. "Slate Roofing." *Bulletin of the Association for Preservation Technology,* p. 77.

[35] Peterson, *Building in Early America,* p. 138.

[36] McKee, "Slate Roofing," p. 77.

[37] Beverly, *History of Virginia,* p. 290.

[38] Peterson, *Building in Early America,* pp. 138–139.

[39] McKee, "Slate Roofing," pp. 78–79.

[40] *Ibid.*, p. 80.

[41] Peterson, *Building in Early America,* p. 140.

[42] Winston-Salem, Southern Decorative Arts. Index to Early Artists. Harrower, John, 13 May 1774.

[43] *Ibid., American and Commercial Daily Advertiser,* 9 August 1806.

[44] Peterson, *Building in Early America,* p. 136.

[45] *Ibid.*

Winston-Salem, Southern Decorative Arts. Index to Early Artists. *The Times,* 7 October 1800.

[46] Peterson, *Building in Early America,* p. 136.

"Comparative Expense of Covering Roofs of Houses." *Journal of the Franklin Institute* 6, no. 4 (October 1830), p. 256.

[47] Peterson, *Building in Early America,* p. 144.

[48] *Ibid.*, pp. 141–142.

[49] Winston-Salem, Southern Decorative Arts. Index to Early Artists. *Federal Gazette and Baltimore Daily Advertiser,* 4 January 1816.

[50] *Ibid., American and Commercial Daily Advertiser,* 18 March 1816.

[51] *Ibid., Richmond Commercial Compiler,* 1 July 1818.

[52] Peterson, *Building in Early America,* p. 142.

[53] *Ibid.*

[54] *Ibid.*

[55] Peterson, *Building in Early America,* p. 41.

[56] Waite, Diana S. *Nineteenth-Century Tin Roofing and Its Use at Hyde Hall.* New York: New York State Historic Trust, 1971, p. 1.

[57] Hoare, W. E.; Hedges, E. S.; Barey, B. T. K. *The Technology of Tinplate.* New York: St. Martins Press, 1965, p. 5.

[58] *Ibid.*, p. 24.

[59] Waite, *Tin Roofing,* p. 2.

[60] Hoare, et al., pp. 12–13.

CHAPTER TWO: THE DECISION TO SPECIALIZE

[1] Bridenbaugh, Carl. *The Colonial Craftsman.* New York: New York University Press, 1950, p.4.

[2] Waldo, Arthur L. *True Heroes of Jamestown.* Miami: American Institute of Polish Culture, 1977, p. 78.

[3] Waldo, *Jamestown,* p. 101.

[4] *Ibid.*

[5] *Ibid.*, pp. 15, 20–21.

[6] Bridenbaugh, p. 27.

[7] *Ibid.*, p. 38.

[8] *Ibid.*, pp. 65, 75.

[9] Wiffen, Marcus. *The Public Buildings of Williamsburg: Colonial Capital of Virginia.* Williamsburg: Colonial Williamsburg, Inc., 1958, p. 231.

[10] Ericsson, Henry. *Sixty Years a Builder: The Autobiography of Henry Ericsson.* Chicago: A. Kroch & Son, 1942, pp. 151–152.

[11] *Ibid.*, p. 187.

[12] Winston-Salem, North Carolina, Museum of Early Southern Decorative Arts. Index of Early Southern Artists and Artisans. *Colonial Mirror and Alexandria Advertiser,* 19 April 1794.

[13] *The Maryland Gazette: A Baltimore General Advertiser,* 11 March 1785.

Federal Gazette and Baltimore Daily Advertiser, 29 April 1802.

American and Commercial Daily Advertiser [Baltimore], 19 June 1815.

Federal Gazette and Baltimore Daily Advertiser, 30 May 1804.

American and Commercial Daily Advertiser, 26 April 1814.

The Maryland Journal and Baltimore Advertiser, 27 October 1789.

The Maryland Journal and Baltimore Advertiser, 12 February 1793.

Federal Gazette and Baltimore Daily Advertiser, 5 October 1802.

American and Commercial Daily Advertiser, 12 August 1820.

[14] *American and Commercial Daily Advertiser,* 12 August 1820.

[15] *Baltimore Directory for 1822 and 1823*. Baltimore: 1822, p. 227.

[16] *Ibid*.

[17] Fitch, James M. *American Building*. New York: Shocken Books, 1966, p. 21.

[18] Waite, Diana S., ed. *Architectural Elements: The Technological Revolution*. Princeton: The Pyne Press, n.d., p. 6.

[19] Stimpson, Charles. *Stimpson's Boston Directory*. Boston: By the author, 1842, p. 371.

Stimpson, Charles. *Stimpson's Boston Directory*. Boston: By the author, 1843, p. 386.

Stimpson, Charles. *Stimpson's Boston Directory*. Boston: By the author, 1844, p. 415.

Stimpson, Charles. *Stimpson's Boston Directory*. Boston: By the author, 1845, p. 389.

Stimpson, Charles. *Stimpson's Boston Directory*. Boston: By the author, 1846, p. 420.

French, James and Stimpson, Charles. *Adams Boston Directory*. Boston: By the authors, 1847, p. 170.

[20] Adams, George. *The Boston Directory*. Boston: James French and Charles Stimpson, 1848, p. 267.

Adams, George. *The Boston Directory*. Boston: By the author, 1849, p. 288.

Adams, George. *The Boston Directory*. Boston: By the author, 1850, p. 324.

Adams, George. *The Boston Directory*. Boston: By the author, 1851, p. 259.

Adams, George. *The Boston Directory*. Boston: By the author, 1852, p. 197.

Adams, George. *The Boston Directory*. Boston: By the author, 1853, p. 290.

Adams, George. *The Boston Directory*. Boston: By the author, 1854, p. 318.

Adams, George. *The Boston Directory*. Boston: By the author, 1855, p. 315.

[21] Stimpson, Charles. *Stimpson's Boston Directory*. Boston: by the author, 1846, p. 215.

French, James and Stimpson, Charles. *Adams's Boston Directory*. Boston: By the authors, 1847, p. 99.

Adams, George. *The Boston Directory*. Boston: James French and Charles Stimpson, 1848, pp. 125, 232.

Adams, George. *The Boston Directory*. Boston: By the author, 1849, p. 130.

Adams, George. *The Boston Directory*. Boston: By the author, 1852, p. 225.

[22] Desilver, Robert. *Desilver's Philadelphia Directory*. Philadelphia: By the author, 1835, p. 55.

Desilver, Robert. *Desilver's Philadelphia Directory*. Philadelphia: By the author, 1837, p. 69.

McElroy, A. *A. McElroy's Philadelphia Directory*. Philadelphia: By the author, 1837, p. 111.

Rogers, Orrin. *A. McElroy's Philadelphia Directory*. Philadelphia: By the author, 1841, pp. 58, 135.

Rogers, Orrin. *A. McElroy's Philadelphia Directory*. Philadelphia: By the author, 1842, p. 58.

Biddle, Edward C. and Biddle, John. *McElroy's Philadelphia Directory*. Philadelphia: By the authors, 1855, p. 273.

Biddle, Edward C. and Biddle, John. *McElroy's Philadelphia Directory*. Philadelphia: By the authors, 1856, p. 320.

[23] Boyd, William. *Boyd's Philadelphia City Business Directory*. Philadelphia: T. K. Collins, Jr., 1858, p. 239.

Stimpson, Charles. *Stimpson's Boston Directory*. Boston: By the author, 1845, p. 72.

[24] Weeks, J. W. *Detroit City Directory*. Detroit: By the author, 1880, p. 244.

[25] Weeks, J. W. *Detroit City Directory*. Detroit: By the author, 1882, p. 128.

[26] Weeks, J. W. *Detroit City Directory*. Detroit: By the author, 1884, p. 336.

CHAPTER THREE: NECESSITY BREEDS INVENTION

[1] Davis, Betsy Warren. *The Warren, Jackson and Allied Families*. Philadelphia: J. B. Lippincott Co., 1903, p. 26.

[2] *Ibid*.

[3] *Ibid*.

[4] Jakle, John A. *Images of the Ohio Valley: A Historical Geography of Travel, 1740–1860*. New York: Oxford University Press, 1977, p. 147.

[5] Cist, Charles. *Sketches and Statistics of Cincinnati in 1859*. Cincinnati: By the author, 1859, pp. 354–355.

[6] Sharrer, G. Terry. "Naval Stores, 1781–1881." *Material Culture of the Wooden Age*. Edited by Brooke Hindle. Tarrytown, N.Y.: Sleepy Hollow Press, 1981, p. 261.

[7] Davis, *Warren Family*, pp. 21, 35, 37.

[8] Haynes, Williams. *American Chemical Industry: Background and Beginnings*, 3 vols. New York: Van Nostrand, 1954, p. 315.

[9] Warren, George Copp. *The Part of the Warrens in the Development of Coal Tar, Petroleum Oil, and Asphalt*. n.p.: Warren Brothers, 1928, p. 5.

[10] *Ibid*., p. 6.

[11] Haynes, *American Chemical Industry*, p. 314–315.

[12] Warren, *Warrens in the Development of Coal Tar*, p. 6.

[13] Black, Jan Knippers, et. al. *Area Handbook for Trinidad and Tobago*. Washington, D.C.: U. S. Government Printing Office, 1976, p. 13.

[14] Peterson, Charles E. "California Brea Roofs." *Master Key* 39, no. 3 (July–September 1965), p. 115.

[15] Warren, *Warrens in the Development of Coal Tar*, p. 8.

[16] *Ibid*., p. 12.

[17] Master Builders Exchange. *Handbook and Catalogue*. Philadelphia: By author, 1895.

[18] Haynes, *American Chemical Industry*, p. 315.

[19] Block, Marguerite Beck. *The New Church in the New World: A Study of Swedenborgianism in America*. New York: Octagon Books, Inc., 1968, p. 101.

[20] Smith, Ophia D. "The New Jerusalem Church in Ohio from 1848 to 1870." *Ohio State Archaeological and Historical Quarterly* 62, no. 1 (January, 1953), p. 43.

[21] Warren, *Warrens in the Development of Coal Tar*, p.5.

[22] Smith, "The New Jerusalem," p. 43.

[23] Warren, *Warrens in the Development of Coal Tar*, p. 5.

[24] Haynes, *American Chemical Industry*, p. 315.

[25] Block, *The New Church*, p. 121.

[26] Unonius, Gustaf. *A Pioneer in Northwest America, 1841–1858*. Minneapolis: Swedish Pioneer Historical Society, 1861; reprinted, 1950, pp. 174–188.

[27] Goodspeed Company. *Industrial Chicago: The Building Interests Vol. I*. Chicago: Goodspeed Company, 1891, p. 350.

[28] Barrett Division. *One Hundred Years: The Barrett Trail*. East Orange, N. J.: By the author, 1954, p. 1.

[29] Illinois Commandery. *Memorials of Deceased Companions of the Commandery of the State of Illinois Military Order of the Loyal Legion of the United States*. Chicago: Illinois Commandery, 1923, p. 91.

[30] Barrett Division, *One Hundred Years*, p. 3.

[31] Haynes, *American Chemical Industry*, p. 315.

[32] Bailey, S. *Bailey's Directory of Chicago*. Chicago: By the author, 1862.

[33] Smith, P. and P. DuMoulin. *A Directory of Chicago*. Chicago: By the authors, 1859.

[34] Bearss, Edwin C. "Pvt. Charles E. Affeld Describes The Mechanicsburg Expeditions." *Journal of the Illinois State Historical Society* 56 (Summer 1963), p. 238.

[35] Illinois Commandery, *Memorials*, p. 91.

[36] Guyer, Isaac D. *History of Chicago: Its Commercial and Manufacturing Interests*. Chicago: 1862, p. 118.

37 Barrett Division, *One Hundred Years,* p. 3.

38 Fite, Gilbert. *The Farmer's Frontier*. New York: Holt-Rhinehart, 1966, p. 73.

39 Barrett Division, *One Hundred Years,* pp. 4–5.

40 Davis, *Warren Family,* p. 35.

41 Biddle, Edward C., and John. *McElroy's Philadelphia Directory for 1853*. Philadelphia: By the authors, 1853.

42 Davis, *Warren Family,* p. 40.

43 Biddle, Edward C., and John. *McElroy's Philadelphia Directory for 1850*. Philadelphia: By the authors, 1853.

44 *American Roofer,* November 1937, p. 12.

45 *United Roofer,* December 1944, p. 14.

46 Irvine, William. *Warren-Ehret-Linck: Case Study*. Chicago: National Roofing Contractors Association, 1984.

47 *Ibid.*

48 *Ibid.*

49 *Ibid.*

50 *Ibid.*

51 *Ibid.*

CHAPTER FOUR: MAN AND HIS MACHINES

1 Lake Side Publishing Company. *The Lake Side Directory of Chicago*. Chicago: By the author, 1877.

2 Aeroil Products Company. *Sixtieth Anniversary Salute to Roofing*. South Hackensack, N.J.: By the author, 1977.

3 Chandler, Alfred D. "The Beginnings of Big Business in American Industry." *Business History Review* 33 (Spring), 1–31.

4 Brody, David. *Essays on the Age of Enterprise: 1870–1900*. Hinsdale, Illinois: The Dryden Press, 1974.

5 *Ibid.*

6 Hays, Samuel P. *The Response to Industrialism: 1885–1914*. Chicago: University of Chicago Press, 1957, p. 36.

7 Wiebe, Robert H. *The Search for Order, 1877–1920*. New York: Hill and Wang, 1967.

8 *Ibid.*

9 Mayer and Wade, *Growth of a Metropolis*. Chicago: University of Chicago Press, 1969.

10 Goodspeed Company. *Industrial Chicago: The Building Interests Volume I*. Chicago: By the author, 1891, p. 343.

11 *Ibid.*, p. 354.

12 Lake Side Press. *Reminiscences of Chicago During the Great Fire*. Chicago: By the author, 1915, p. 6.

13 *Ibid.*, pp. 94–95.

14 *Ibid.*

15 *Ibid.*, p. 23.

16 *Ibid.*, pp. 40–41.

17 *Ibid.*, p. xxvi.

18 *Ibid.*, pp. xxviii, xxix.

19 *Ibid.*, p. 86.

20 Baker, Maxwell C. *Roofing-Past and Present*. Ottawa: National Research Council of Canada, 1965, p. 3.

21 Sheetmetal Roofing Trade. *Selling Arguments for Tin Roofing*. Philadelphia: By the author, 1911, p. 30.
National Board of Fire Underwriters. *Shingle Roofs as Conflagration Spreaders: An Appeal to the Civil Authorities and Civic and Commercial Bodies*. New York: By the author, 1916, p. 1.

22 *Ibid.*

23 *Ibid.*, p. 3.

24 *Ibid.*

25 *Ibid.*, p. 15.

[26] Journal of the Franklin Institute. "Comparative Expense of Covering Roofs of House." *Journal of the Franklin Institute* 6, no. 4 (October 1830), p. 255.

[27] Freitag, Joseph K. *Fire Prevention and Fire Protection as Applied to Building Construction: A Handbook of Theory and Practice.* New York: John Wiley and Sons, Inc., 1921, p. 681.

[28] National Board of Fire Underwriters, *Shingle Roofs,* p. 21.

[29] Freitag, *Fire Prevention,* p. 680.

[30] *Ibid.*

[31] Richardson Roofing/Flintkote Company. *Richardson Asphalt Products.* New York: By the author, 1934, p. 7.

[32] *Ibid.*

[33] American School of Correspondence. *Cyclopedia of Fire Prevention and Insurance,* 4 volumes. Chicago: By the author, 1912, pp. 103, 209.

[34] National Board of Fire Underwriters, *Shingle Roofs,* p. 16.

[35] Moore, Frances C. *Fire Insurance and How to Build.* New York: John Wiley and Sons, 1903, p. 645.

[36] The Chronicle Company. *The Chronicle Fire Tables for 1894.* New York: By the author, 1894, p. 4.

The Chronicle Company. *The Chronicle Fire Tables for 1896.* New York: By the author, 1896, p. 2.

The Chronicle Company. *The Chronicle Fire Tables for 1900.* New York: By the author, 1900, p. 6.

The Chronicle Company, *The Chronicle Fire Tables for 1903.* New York: By the author, 1903, p. 6.

[37] H. W. Johns. *Asbestos Roofing and Other Materials Manufactured by H. W. Johns.* New York: By the author, 1872, p. 17.

B. F. Barrett. *Superior Fire and Waterproof Roofs and Roofing Materials, Manufactured by B. F. Barrett and Co.* Chicago: By the author, 1853, pp. 1–2.

[38] H. W. Johns, *Asbestos Roofing,* p. 17.

[39] Waite, Diana S. *19th Century Tin Roofing and Its Use at Hyde Hall.* New York: New York State Historical Trust, 1971, p. 305.

[40] Sheetmetal Roofing Trade. *Selling Arguments for Tin Roofing.* Philadelphia: By the author, 1911, p. 34.

[41] *Ibid.*, p. 37.

[42] Edwards Manufacturing Company. *World's Best Roofing.* Cincinnati: By the author, 1916, p. 12.

[43] *Ibid.*

[44] Hoare, W. E.; Hedges, E. S.; Barry, B. T. K. *The Technology of Tinplate.* New York: St. Martins Press, 1965, pp. 13, 24, 25.

[45] Clark, Victor S. *History of Manufacturers in the United States: Volume 2 1860–1893.* New York: Carnegie Institution of Washington, 1929, p. 99.

[46] *Ibid.*, p. 373.

[47] Hoare, et. al., *Technology of Tinplate,* p. 8.

[48] Clark, *History of Manufacturers,* vol. 2, p. 373.

[49] *Ibid.*, p. 133.

[50] *Ibid.*, pp. 130–131.

[51] Journal of the Franklin Institute. "Truman's Made of Rendering Tin Plate Durable." *Journal of the Franklin Institute* 4, no. 3 (March 1832), p. 179.

[52] Waite, *19th Century Tin Roofing,* p. 6.

[53] Waite, *19th Century Tin Roofing,* p. 9.

[54] Clark, *History of Manufacturers,* vol. 2, p. 375.

[55] Waite, Diana S., ed. *Architectural Elements: The Technological Revolution.* Princeton: The Pyne Press, n.d., p. 7.

[56] Paul, Rodman Wilson. *Mining Frontiers of the West 1848–1880.* New York: Holt, Rinehart and Winston, 1963.

[57] Waite, *Architectural Elements,* p. 8.

[58] Clark, *History of Manufacturers,* vol. 2, pp. 97, 398.

[59] *Ibid.*, p. 367.

60 *Ibid.*, p. 368.

61 John Hamilton. *Practical and Ornamental Metal Roofing*. Pittsburgh: By the author, 1896, pp. 7–8.

62 N. Y. Slate Roofing Company. *Glines' Patent Slate Roofing Paint*. New York: By the author, 1874, pp. 7–8.

63 Barrett Manufacturing Company. "A Short History of the Bonded Roof." East Orange, N.J.: By the author, 1949, p. 1.

64 *Ibid*.

65 Mica Roofing Company. *Mica Roofing Company, Manufacturers and Dealers in Composition Roofing, etc.* New York: By the author, 1869, p. 7.

66 H. W. Johns, *Asbestos Roofing*, p. 2.

67 The Phillip Carey Manufacturing Co. *The Carey Roof*. Cincinnati: By the author, 1904, pp. 17, 37.

68 Edwards Manufacturing Company, *World's Best Roofing*, p. 14.

69 Pendergast, James B. *This Problem Could Set the Roofing Industry Back 50 Years*. New York: Allied Chemical, Barrett Division, n.d., p. 1.

70 Barrett Manufacturing Company, *Short History of the Bonded Roof*, pp. 1–2.

71 Wren, Douglas. "A History of Waterproof Coatings for Roofs." *Bulletin Technique*. Quebec: Oudue des Architectes due Quebec, p. 4.

72 Schreiber, Edward T. "Roof Failures: Whose Responsibility?" *Architectural and Engineering News* (November 1964), pp. 56–57.

73 Pendergast, *This Problem*, p. 2.

74 Johns-Manville Corporation. "Built-up Roofing." *Architectural Service Bulletin* No. 1. (1929), p. 5.

75 Richardson Roofing/Flintkote Company, *Richardson Asphalt Products*, p. 6.

76 *Ibid*.

77 United States Department of Commerce. *Historical Statistics of the United States: Colonial Times to 1970*, 2 volumes. Washington, DC: U.S. Government Printing Office, 1975, p. 623.

78 Barrett Division, *One Hundred Years: The Barrett Trail*, p. 3.

79 Polk, Ralph L. *Detroit City Directory for the Year Commencing August 1, 1893*. Detroit: By the author, 1893, p. 32.

80 Glenhead, N.Y., Records and Archives of Nicholson and Galloway, Inc., 1900.

81 Phillips, Joseph D. *Little Business in the American Economy*. Urbana: University of Illinois Press, 1958, p. 52.

82 U.S. Department of Commerce, *Historical Statistics of the United States*, p. 142.

CHAPTER FIVE: ECONOMIC EXPANSION

1 Allen, Frederick Lewis. *The Big Change: America Transforms Itself, 1850–1900*. New York: Bantam Books, 1961, p. 164.

2 Aeroil Products Company. *Sixtieth Anniversary Salute to the Roofing Industry*. South Hackensack, N.J.: Privately Printed, 1977, pp. 1, 4.

3 Wehner Roofing and Tinning Company. *The Same Family has Served Dayton for 129 Years*. Dayton: Privately Printed, 1797.

4 Powell, M. W., Business Card, M. W. Powell Company Records, Chicago, Ill. 1923.

5 Board of Directors, February, 1925. J. D. Candler Roofing Company Records, Detroit, Mich.

6 Aeroil, *Salute to the Roofing Industry*, p. 14.

7 U.S. Department of Commerce, *Historical Statistics of the United States: Colonial Times to 1970*, p. 623.

8 Long, Clarence D. "Long Cycles in the Building Industry." *Quarterly Journal of Economics* 53, no. 3 (May 1939), p. 398.

9 Phillips, Joseph D. *Little Business in the American Economy*. Urbana: University of Illinois Press, 1958, p. 52.

10 *Ibid.*, p. 142.

11 Aeroil, *Salute to the Roofing Industry*, p. 5.

12 *Ibid*.

REFERENCES

13 Irvine, William. *Warren-Ehret-Linck Case Study*.

14 *Ibid*.

15 Irvine, William. *M. W. Powell Company: Case Study*. Chicago: National Roofing Contractors Association, 1984, p. 10.

16 Board of Directors, February 18, 1922. J. D. Candler Roofing Company, Inc. Detroit, Mich.

17 Board of Directors, February 22, 1925. J. D. Candler.

Board of Directors, November 11, 1927. J. D. Candler.

18 *American Roofer* (May 1936), p. 42.

19 *Ibid*.

20 Bernstein, Irving. *The Lean Years: A History of the American Worker, 1920–1933*. Baltimore: Penguin Books, 1960, p. 42.

21 Sachs, Bernard. "Looking Backward: The Birth and Growth of a Great Industry Are Reflected in the Pages of American Roofer." *American Roofer* (January 1940), p. 41.

22 Conlin, Joseph. *The American Past: Part Two—A Survey of American History Since 1865*. New York: Harcourt Brace Jovanovich, 1984, p. 550.

23 Haber, William. *Industrial Relations in the Building Industry*. Cambridge: Harvard University Press, 1930, p. 4.

24 Rand School of Social Science. *The American Labor Year Book—1926*. New York: Rand School of Social Science, 1926, p. 87.

25 Haber, *Industrial Relations*, p. 99.

26 *Ibid*., p. 259.

27 Hurley, J. T. "Report of Organizer." *The Journeymen Roofers Magazine* 1, no. 10 (January 1928), p. 4.

28 Hurley, J. T. "Report of Organizer." *The Journeymen Roofers Magazine* 1, no. 5 (October 1926), p. 4.

29 *Ibid*.

30 Hurley, J. T. "Report of Organizer." *The Journeymen Roofers Magazine* 1, no. 16 (July 1929), p. 2.

31 Hurley, J. T. "Report of Organizer." *The Journeymen Roofers Magazine* 1, no. 7 (April 1927), p. 4–5.

32 Hurley, J. T. "Report of Organizer." *The Journeymen Roofers Magazine* 1, no. 11 (April 1928), p. 6.

33 Hurley, J. T. "Report of Organizer." *The Journeymen Roofers Magazine* 1, No. 12 (July 1928), p. 15.

34 *Ibid*.

35 *Ibid*.

36 *The Journeymen Roofers Magazine* (October 1926), pp. 13–14.

37 Jahoda, Gloria. *Florida: A Bicentennial History*. New York: Norton & Company, 1976, p. 124–125.

38 Hurley, J. T. "Report of Organizer." *The Journeymen Roofers Magazine* 1, no. 5 (October 1926), p. 13.

39 Hurley, J. T. "Report of Organizer." *The Journeymen Roofers Magazine* 1, no. 13 (October 1928), p. 4.

40 *The Journeymen Roofers Magazine* (April 1928), p. 7.

41 Wood, Roderick A. ''30 Years of Insulation." *American Roofer* (January 1940), p. 35.

42 *Ibid*.

43 Tebeau, Charlton W. *A History of Florida*. Coral Gables, Florida: University of Miami, 1971, p. 374.

44 National Roofer. "Death Ends Colorful Career of Bror Dahlberg." *National Roofer* (November 1954), p. 17.

45 Aeroil, *Salute to the Roofing Industry*, p. 1.

46 Prestler, C. F. "I Cut the First Asphalt Shingles Ever Made." *American Roofer* (February 1940), p. 17.

47 VonSternberg, Jule Robert. "The Asphalt Shingle—A Native American." *American Roofer* (January 1940), pp. 42–44.

48 Sachs, "Looking Backward," p. 12.

49 *Ibid*., p. 31.

50 *Ibid*., p. 12.

CHAPTER SIX: COPING WITH HARD TIMES

[1] Conlin, Joseph. *The American Past: Part Two—A Survey of American History Since 1865.* New York: Harcourt Brace Jovanovich, 1984, p. 700.

[2] *Ibid.*

[3] Certainteed Corporation. *Certainteed's First Seventy-Five Years: Sharing in America's Growth.* Valley Forge, VA: Certainteed Corporation, 1978, p. 8.

[4] Conlin, *the American Past,* p. 700.

[5] *Ibid.*, p. 698.

[6] *Ibid.*

[7] *The Journeymen Roofers Magazine* Vol. 2 No. 3, July 1931.

[8] *Ibid.*

[9] *The Journeymen Roofers Magazine* Vol. 1 No. 19 (April 1930), pp. 5–6.

[10] United States Department of Commerce. *Historical Statistics of the United States: Colonial Times to 1970,* 2 vols. Washington: U. S. Department of Commerce, 1975, p. 623.

[11] Phillips, Joseph D. *Little Business in the American Economy.* Urbana: University of Illinois Press, 1958, p. 52.

[12] Sachs, Bernard. "Roofers Battle 'Muscling In' by Lumber Dealers." *American Roofer* (April 1940), pp. 31–32.

[13] Certainteed, *First Seventy-Five Years,* p. 8.

[14] Irvine, William. *M. W. Powell Company: Case Study.* Chicago: M. W. Powell Records, 1984, p. 10.

[15] Irvine, William. *Warren-Ehret-Linck: Case Study.* Chicago: National Roofing Contractors Association, 1984.

[16] *Ibid.*

[17] Conlin, *The American Past,* p. 698.

[18] *American Roofer* (January 1940), p. 20.

[19] Board of Directors, March 26. 1932.

Records and Archives of J. D. Candler, Detroit, Mich.

[20] Board of Directors, June 17, 1932.

Records and Archives of J. D. Candler, Detroit, Mich.

[21] Interview with Frank Lawson, Sr., San Francisco, Calif., May, 1984.

[22] *American Roofer* (June 1937), pp. 18–19.

[23] Certainteed, *First Seventy-Five Years,* p. 9.

[24] McCawley, James. "Asbestos—the Magic Minera." *American Roofer* (April 1937), p. 19.

[25] Blum, John M., et. al. *The National Experience: Part Two—A History of the United States Since 1865.* New York: Harcourt Brace Jovanovich, 1985, p. 677.

[26] Sachs, Bernard. "Looking Backward: The Birth and Growth of a Great Industry Are Reflected in the Pages of American Roofer." *American Roofer* (January 1940), p. 32.

[27] *American Roofer* (January 1962), p. 17.

[28] *American Building Contractor* (April 1969), pp. 16–17.

[29] Interview with Tom Daly, Kansas City, Kan. August 8, 1965.

[30] *Ibid.*

[31] *American Building Contractor* (April 1969), p. 17.

[32] Interview with Tom Daly, August 8, 1985.

[33] *Ibid.*

[34] Phillips, *Little Business,* p. 52.

[35] *American Roofer* (June 1937), pp. 18–19.

[36] Board of Directors, April 20, 1935, J. D. Candler.
Board of Directors, October 19, 1935, J. D. Candler.
Board of Directors, April 27, 1936, J. D. Candler.

[37] Board of Directors, January 1, 1936, J. D. Candler.
Board of Directors, December 12, 1936, J. D. Candler.

[38] Irvine, *Warren-Ehret-Linck: Case Study.*

[39] Smith, Robert B. "Roofing Revival: A Major Item in Federal Housing Administration." *American Roofer* (January 1936).

[40] *American Roofer* (October 1938), p. 13.

[41] Glenhead, N.Y.: Records and Archives of Nicholson and Gallaway, Inc.

[42] Irvine, *M. W. Powell: Case Study*, p. 10.

[43] *Ibid*.

[44] Irvine, *Warren-Ehret-Linck: Case Study*.

[45] Board of Directors, February 19, 1938, J. D. Candler.

[46] Board of Directors, March 19, 1938, J. D. Candler.
Board of Directors, April 16, 1938, J. D. Candler.

[47] Fitch, James Marston. *American Building: The Historical Forces That Shaped It*. New York: Schocken Books, 1966, p. 186.

[48] Bartlett, Larry. "Million Dollar Baby." *American Roofer* (January 1940), p. 24.

[49] *Ibid*.

[50] Mercer, Edgar P. "Roofing Wages and Hours." *American Roofer* (December 1940), p. 6.

[51] *American Roofer* (January 1937), p. 21.

[52] "Roofers Employed on W.P.A. Projects." *American Roofer* (October 1936), p. 30.

[53] "W.P.A. Still Competes with Contractors." *United Roofer* (March 1940), p. 11.

[54] *United Roofer* (March 1940), p. 30.

[55] Brogan, D.W. *Roosevelt and the New Deal*. London: Oxford University Press, 1952, p. 174.

[56] *American Roofer* (September 1938), p. 23.

[57] Smith, "Roofing Revival," p. 6.

[58] Fitzpatrick, Charles P. "Fair Weather *or* Foul—Mr. President." *American Roofer* (November 1937), pp. 7–8.

[59] Sachs, Bernard. "Census Shows Roofing Trends." *American Roofer* (January 1939), p. 7.

CHAPTER SEVEN: CHALLENGES AND OPPORTUNITIES

[1] Sachs, Bernard. "How Will the War Affect Roofing?" *American Roofer* (October 1939), pp. 9–10.

[2] Sachs, Bernard. "Roofers Battle 'Muscling In' by Lumber Dealers." *American Roofer* (April 1940), pp. 9, 37–38.

[3] Tuscany, Arthur J. "Roofing at the Crosswords" *American Roofer* (March and April 1940), pp. 15, 30.

[4] Sachs, Bernard. "How Will the War Affect Roofing?" p. 24.

[5] *Ibid*., p. 10.

[6] Conlin, Joseph. *The American Past: A Survey of American History Since 1865*. New York: Harcourt Brace Jovanovich, 1984, p. 737.

[7] *The Journeyman Roofers Magazine* Vol. 3, No. 3, 3rd Quarter 1943.

[8] *Ibid*.

[9] *Ibid*.

[10] *American Roofer* (August 1943), p. 28.

[11] Wright, W. R. "Roofers Robbed." *United Roofer* (May 1943), p. 11.

[12] Board of Directors, February 24, 1940.
J. D. Candler Roofing Co., Inc., Detroit, Mich.
Board of Directors, September 2, 1942. J. D. Candler.
Board of Directors, April 24, 1943. J. D. Candler.
Board of Directors, January 19, 1945. J. D. Candler.

[13] Interview with Frank Lawson, May, 1984, San Francisco, Calif.

[14] *American Roofer* (February 1943), pp. 14–15.

[15] *American Roofer* (August 1943), pp. 10.

[16] Dunham, C. J. "Delivering the Goods in Wartime." *American Roofer* (June 1944), p. 10.

[17] *American Roofer* (December 1942), p. 13.

[18] Wright, "Roofers Robbed," p. 15.

[19] Dunham, C. J. "Delivering the Goods in Wartime." *American Roofer* (June 1944), p. 10.

[20] *American Roofer* (July 1960), pp. 22–23.

[21] *United Roofer* (March 1941), p. 11.

[22] Carpenter, Milton. "All Roads Lead to Chicago." *United Roofer* (February 1944), p. 4.

[23] *Ibid.*, p. 7.

[24] *United Roofer* (October 1942), p. 16–17.

[25] VonSternberg, Jule Robert. "The Asphalt Shingle—A Native American." *American Roofer* (January 1940), p. 22.

[26] McPherson, C. M. "Asphalt Contractors Set up a National War Council." *American Roofer* (December 1942), p. 18.

[27] Wright, "Roofers Robbed," p. 9.

[28] Mann, William J. "This Aircraft Plant Was Roofed in a Day." *American Roofer* (July 1941), p. 11.

[29] Brown, George. "The Apprentice Problem." *American Roofer* (June 1937), p. 15.

[30] Blake, E. H. "Roofing is Big Business." *American Roofer* (October 1941), p. 15.

[31] Neulander, Alfred E. "Let's Trowel-Coat Hitler." *American Roofer* (June 1943), p.1.

CHAPTER EIGHT: COMING OF AGE

[1] *United Roofer* (January 1945), p. 11.

[2] *United Roofer* (February 1945), p. 11.

[3] Interviews held in Chicago, Ill. with past presidents of NRCA, various dates.

[4] *United Roofer* (August 1945), p. 24.

[5] Christiansen, F. J. A. Roofing Co. *Highlights of 100 Years in the Life of F. J. A. Christiansen Roofing Co., Inc., 1879–1979*. Milwaukee: By the author, 1979, p. 4.

[6] *United Roofer* (December 1945), p. 24.

[7] *United Roofer* (September 1945), p. 24.

[8] United States Department of Commerce. *Historical Statistics of the United States: Colonial Times to 1970,* 2 volumes. Washington, D.C.: U.S. Government Printing Office, 1975, p. 623.

[9] *United Roofer* (September 1945), p. 25.

[10] *United Roofer* (February 1946), p. 9.

[11] McCawley, James. "The Mystery of the Disappearing Felt." *United Roofer* 8, no. 3 (March 1946), pp. 16–17.

[12] *United Roofer* (May 1946), p. 28.

[13] *Ibid.*, p. 12.

[14] Phillips, *Little Businesses,* p. 54.

[15] U.S. Department of Commerce, *Historical Statistics,* p. 142.

[16] Board of Directors. December 20, 1946. J. D. Candler Roofing Co., Inc., Detroit, Mich.

Irvine, William. *Warren-Ehret-Linck: Case Study*. Chicago: National Roofing Contractors Association, 1984.

[17] Fitzpatrick, Charles P. "You Can Write a Sales Letter." *United Roofer* 7, no. 3 (May 1945), p. 10.

[18] Glenhead, N.Y. Records and Archives of Nicholson and Galloway, Inc.

[19] Christiansen, F. J. A. Roofing Co. *Highlights of 100 Years in the Life of F. J. A. Christiansen Roofing Co., Inc., 1879–1979*. Milwaukee: By the author, 1979, p. 4.

[20] *United Roofer* (October 1947), p. 17.

[21] Cockrell, Charles L. "Labor Relations in the Roofing, Siding and Insulation Industry." *United Roofer* 10, no. 11 (November 1948), p. 16.

[22] *National Roofer* (May 1949), pp. 4, 46.

[23] *National Roofer* (March 1953), p. 3.

[24] Conlin, Joseph. *The American Past: A Survey of American History Since 1865*. New York: Harcourt Brace Jovanovich, 1984, p. 779.

[25] Smith, F. T., "Application Goes Modern." *American Roofer* (January 1940), p. 28.

[26] Sachs, Bernard, ed. *The Built-up Roof*. New York: Harris-Fox-Hoffman Corp., 1939, pp. 14–15.

[27] *United Roofer* (August 1945), p. 18.

[28] *United Roofer* (August 1947), p. 27.

[29] "New Slag Scraper." *United Roofer* (September 1947), p. 17.

[30] *United Roofer* (January 1948), p. 14.

[31] *National Roofer* (July 1955), p. 3.

[32] *United Roofer* (June 1948), p. 38.
National Roofer (March 1949), p. 2.

[33] *United Roofer* (April 1948), p. 36.

[34] *American Roofer* (April 1969), p. 18.

[35] *Ibid.*, pp. 18, 30.

[36] Interview with Fred Good, Chicago, Ill. December 3, 1983.

[37] NRCA Past Presidents' Interviews.

[38] *Ibid*.
National Roofer (March 1957), p. 4.

[39] Interview with Fred Good.

[40] U.S. Subcommittee on Antitrust and Monopoly of the Committee on the Judiciary, *Senate Report with Individual Views*. Washington, D.C.: U.S. Government Printing Office, 1958.

[41] *Ibid.*, pp. 5–6.

[42] *Ibid.*, p. 6.

[43] *Ibid.*, pp. 9–10.

[44] *Ibid.*, p. 39.

[45] Interview with Edward T. Schreiber, Detroit, Mich. July 7, 1983.

[46] *Ibid*.

[47] Cassimatis, Peter J. *Economics of the Construction Industry*, Publication III. New York: National Industrial Conference Board, n.d., pp. 11–14.

[48] U.S. Department of Commerce, *Historical Statistics of the United States*, p. 142.

[49] Irvine, William. *Warren-Ehret-Linck: Case Study*. Chicago: National Roofing Contractors Association, 1984.
Board of Directors, November 1, 1967. J. D. Candler.

[50] NRCA Past Presidents' Interviews.

[51] Interview with Fred Good.

[52] NRCA Past Presidents' Interviews.

CHAPTER NINE: LESSONS OF THE PAST

[1] *Roofing Spec* (April 1967), p. 4.

[2] *Roofing Spec* (September 1967), p. 9.

[3] *Roofing Spec* (March 1971), p. 19.

[4] *American Roofer* (January 1940), p. 20.

[5] *Roofing Spec* (November 1971), p. 21.

[6] *Roofing Spec* (May 1973), p. 12.

[7] *Roofing Spec* (May 1972), p. 9.

[8] *Roofing Spec* (November 1972), p. 4.

[9] Stender, John H. "OSHA's Target Industry Program—A Challenge." *Roofing Spec* 2 (May 1974), p. 32.

[10] *Roofing Spec* (November 1978), p. 17.

[11] *Ibid*.
Roofing Spec (September 1979), p. 12.

[12] *Roofing Spec* (March 1976), pp. 11–13.

[13] Interviews with past presidents of NRCA, Chicago, various dates.

[14] *Ibid*.

[15] *Roofing Spec* (September 1975), p. 10.

[16] *Roofing Spec* (July 1976), pp. 14–15.
Roofing Spec (May 1971), pp. 21–22.

[17] *Roofing Spec* (January 1973), p. 19.

[18] Roth, Leland M. *A Concise History of American Architecture*. New York: Harper and Row, 1979, p. 295.

[19] *Ibid*.

[20] *Ibid*.

[21] Whiffen, Marcus and Koeper, Frederick. *American Architecture 1607–1976*. Cambridge: MIT Press, 1981.

[22] Personal Communication, Sarnafit, Inc., 1983.

[23] *Ibid*.

[24] *Ibid*.

[25] *Roofing Spec* (May 1980), p. 45.
Roofing Spec (May 1979), pp. 18–19.

[26] *Roofing Spec* (November 1978), p. 68.

[27] *Roofing Spec* (November 1974), pp. 31–33.
"Metal Roofing . . . Versatility and Style," *Roofing Spec* (November 1974), p. 29.

[28] *Ibid*., p. 28.

[29] *Roofing Spec* (October 1975), p. 28.

[30] Fred Good, Interview with T. J. Karamanski, August 18, 1985.

[31] *Roofing Spec* (May 1980), p. 45.

[32] *Roofing Spec* (November 1977), p. 28.

[33] *Roofing Spec* (March 1979), pp. 8–9.

[34] *Roofing Spec* (January 1980), p. 38.

[35] *Roofing Spec* (March–April 1970), pp. 11–12.

[36] *Ibid*., pp. 5–11.

[37] *Wall Street Journal*, July 26, 1978, p. 20.
Good, Bill. "Is Asbestos Dangerous? Johns-Manville Fights Back." *Roofing Spec* (October 1980), p. 35.

[38] *Wall Street Journal* (October 30, 1978), p. 4.

[39] *Ibid*.

[40] *Wall Street Journal* (March 8, 1979), p. 3.
Wall Street Journal (May 7, 1979), p. 4.

[41] *Wall Street Journal* (November 23, 1979), p. 18.

[42] "Jury Rules Against Two Asbestos Makers in Asbestos Lawsuit." *Wall Street Journal* (May 29, 1980), p. 21.

[43] *Wall Street Journal* (June 30, 1980), pp. 1, 10.

[44] Good, "Is Asbestos Dangerous?" pp. 35–36.

[45] *Wall Street Journal* (May 7, 1981), p. 22.
Wall Street Journal (November 24, 1981), p. 38.

[46] *Wall Street Journal* (August 27, 1982), p. 1.
Wall Street Journal (August 30, 1982), p. 15.

[47] *United Roofer* (February 1949), p. 18.

[48] *Roofing Spec* (March 1970), pp. 22–25.

[49] Northrup, Herbert and Howard G. Foster. *Open Shop Construction*. Philadelphia: The Wharton School, University of Pennsylvania Press, 1981, pp. 12–13.

[50] *Ibid*., pp. 195–198.

[51] *Ibid*., p. 19.

[52] *Roofing Spec* (February 1984), pp. 61–64.

"There is properly no history, only biography." RALPH WALDO EMERSON

If to give one's self is the greatest gift of all, then NRCA has been truly blessed with the generosity of its volunteer leaders.

Only someone who has served as president of the Association can understand the level of commitment this position requires. It takes a very special man to willingly, repeatedly turn his attention from a comfortable home and a demanding business to work for which he will receive few thanks and no remuneration.

Of course, we all have our limits. And when these people reached theirs, they asked if there wasn't something more they could do—to raise money for a worthy project, or explain an industry issue to an allied group, or revive a foundering program.

It is a challenge to establish a forum for a tribute befitting these gentlemen. We hope they will live on in our hearts and minds if we commit their accomplishments to these pages.

This information was compiled from biographical forms submitted by past presidents of NRCA and their families.

1882* .... **COL. M. W. POWELL** Chicago, Ill.
1883* .... **H. M. REYNOLDS** Grand Rapids, Mich.
1884* .... **MAJ. SAMUEL E. BARRETT** Chicago, Ill.
1885* .... **SAMUEL D. WARREN** St. Louis, Mo.
1886 .... **J. WILKES FORD** Chicago, Ill.
1887 .... **F. J. SOKUP** Grand Rapids, Mich.
1888 .... **H. R. SHAFFER** Chicago, Ill.
1889 .... **HENRY C. SMITHER** Indianapolis, Ind.
1890 .... **E. S. BORTEL** Philadelphia, Pa.
1891 .... **W. P. LUPTON** Pittsburgh, Pa.
1892 .... **GEORGE WINDING** Milwaukee, Wis.
1893 .... **WILLIAM O'DONNELL** Chicago, Ill.
1894 .... **S. A. DATZELL** Youngstown, Ohio
1895 .... **JOHN M. SELLERS** St. Louis, Mo.
1896 .... **C. G. JAMESON** Buffalo, N.Y.
1897 .... **A. J. SHIRK** Oklahoma City, Okla.
1898 .... **CHARLES KUNZLER** Pittsburgh, Pa.
1899 .... **CHARLES SCHNEIDER** St. Louis, Mo.
1900 .... **JOHN C. FINEGAN** Boston, Mass.
1901 .... **HENRY REUTER** Kankakee, Ill.
1902 .... **W. N. CHILDS** New York, N.Y.
1903 .... **MAJ. ELIAS POWELL** Chicago, Ill.
1904 .... **J. LOBER** Philadelphia, Pa.
1905  **J. W. BUSHNELL** Peoria, Ill.

* *Although the Gravel Roofers Protective Association, the forerunner of NRCA, was officially founded in 1886, a loosely organized group of Chicago roofers had begun electing officers as early as 1882. Between 1882 and 1886, some of the industry's most important leaders served this tiny faction and helped it grow to achieve national status.*

1906 **HOWARD NICHOLS** Des Moines, Iowa

1907 .... **E. W. JAMESON** Buffalo, N.Y.
1908 .... **A. F. GOLDEN** Scranton, Pa.
1909 .... **WILLIAM B. HART** Providence, R.I.
1910 .... **W. L. SPRINGER** Chicago, Ill.
1911 .... **W. A. MURTFELD** Boston, Mass.
1912 .... **JOHN L. JONES** Chicago, Ill.
1913 .... **L. W. HARRINGTON** New York, N.Y.
1914 **LEE H. GOULD** Cleveland, Ohio

1915 .... **JOHN INGRAM** Chicago, Ill.
1916 .... **GEORGE W. MOORE** Providence, R. I.
1917 .... **A. W. BARTHOLOMEW** Memphis, Tenn.
1918 **J. C. NORTON** Cleveland, Ohio

1919–1920 **T. F. SLOAN** Paterson, N.J.

1921 **MARK A. CRONIN** (1888–1956)
Knickerbocker Roofing and Paving Co., Chicago, Ill.
NRCA committees on labor and membership
Spouse: Emily Frances; children: James, Robert, Dorothy, Kathleen, Mark Jr.

1922 **CHARLES N. LOUIS** Peoria, Ill.

1923 **JOHN C. FINEGAN** Boston, Mass.

1924  **A. W. CARMICHAEL** Toronto, Canada

1925  **GEORGE E. MOORE** Chicago, Ill.

1926  **NORMAN E. DOW** Omaha, Neb.

1927 .... **GEORGE E. MOORE** Chicago, Ill.

1928  **C. C. JAMES** Grand Rapids, Mich.

1929 .... **JOSEPH CAMPBELL** Providence, R.I.

1930–1931  **J. BOYD GRIFFITHS** (1892–1966)
Binghamton Slag Roofing Co., Inc.,
Binghamton, N.Y.
NRCA committee member
NRCA J. A. Piper Award 1954
National Recovery Act Committee on
Roofing and Sheet Metal chair, WWII
Spouse: Emily; child: Charles.

1932–1933  **JOSEPH A. PIPER** (1878–1950)
J. A. Piper Roofing Co., Inc.,
Greenville, S.C.
NRCA committee member
NRCA Distinguished Service Award 1947
Namesake of NRCA's J. A. Piper Award 1949
One of the founders of the Carolinas Roofing
and Sheet Metal Contractors Association
1943; helped organize the Georgia Sheet
Metal Contractors Association 1949
Spouse: Mamie; children: Joe, Herbert,
Sam, Rodney.

1934  **C. L. PITTS** Newark, N.J.

1935  **JOHN J. HESSION** Louisville, Ky.

1936

D. A. JACKSON Los Angeles, Calif.

1937–1938

THOMAS J. DALY (1899–)
Kaw Roofing and Sheet Metal Inc., Kansas City, Mo.
NRCA committee member
NRCA J. A. Piper Award 1950
Midwest Roofing Contractors Association James Q. McCawley Award;
National Recovery chair in Kansas in WWII
Spouse: Margaret; children: Edward, Rosemary, Patrick, Kathleen.

1939

IRWIN A. LANGER (1903–1984)
Langer Roofing and Sheet Metal Inc., Milwaukee, Wis.
NRCA committee member
NRCA J. A. Piper Award 1948
Associated Roofing Contractors of Milwaukee secretary-treasurer
Spouse: Jeannette; children: Glenn, Jean, Elizabeth.

1940–1941

MYRON W. POWELL (1891–1960)
M. W. Powell Co., Chicago, Ill.
NRCA committee member
Spouse: Margaret; children: Marion, Lillian.

1942

MILTON CARPENTER (1905–)
Trinidad Asphalt Co., St. Louis, Mo.
NRCA committee member
Doctorate in letters, Washington University
Roofing Contractors of St. Louis president 1939–1940.

1943

H. O. REX Stamford, Conn.

1944–1945 .... **A. BIEBEL** St. Louis, Mo.

1946

C. C. HIGH Scranton, Pa.

1947–1948

C. L. COCKRELL Chicago, Ill.

1949 

BEN J. ESKO (?–1963)
Esko Roofing Co., Chicago, Ill.
NRCA committee member
Spouse: Frances; child: Irwin.

1950 

GORDON M. WATERS (1910–)
Interstate Roofing Consultants, Inc., Tampa, Fla.
NRCA committee member
NRCA J. A. Piper Award 1957
Carolinas Roofing and Sheet Metal Association first president 1943; namesake of Carolinas Association's Gordon M. Waters Award
Spouse: Marijane; children: Gordon Jr., Henry.

1951 .... **MERT T. BUCKLEY** (1893–1977)
Buckley Construction Industries, Wichita, Kan.
NRCA committee member
NRCA J. A. Piper Award 1951
Midwest Roofing Contractors Association co-founder
Spouse: Pearl; children: Duane, William.

1952 

B. D. SCHRAM Denver, Colo.

1953 

WALTER J. SIMON (1901–1976)
St. Clair Roofing Co., St. Louis, Mo.
NRCA committee member
NRCA J. A. Piper Award 1960
Roofing Contractors of St. Louis president
Spouse: Octavia; children: Don, Eugene.

1954–1955 .... **JOHN D. ENNIS** Terre Haute, Ind.

1956 .... **HARVEY WALLACE** New Orleans, La.

1957 .... **CALVIN BOWMAN** (1909–)
Standard Roofing and Material Co., Oklahoma City, Okla.
NRCA Committee on Technical Services and Research
NRCA J. A. Piper Award 1963
Spouse: Opal; children: Carol, Calvin Jr.

1958 

CLYDE H. SCOTT (1907–1974)
Empire Roofing Co., Chicago, Ill.
NRCA committees on labor, technical services and research
NRCA J. A. Piper Award 1961
Chicago Roofing Contractors Association president; namesake of CRCA's Clyde Scott Award
Spouse: Thyra; children: Eugene, Lois.

1959–1960 

JOHN P. REUTER JR. (1899–1977)
Missouri Roofing Co., St. Louis, Mo.
NRCA committee member
NRCA J. A. Piper Award 1962
Roofing Contractors of St. Louis president
Spouse: Helen; children: James, Jane, Jack.

1961

RUDOLPH J. BARNES (1910–1970)
G. G. Ray Co., Charlotte, N.C.
NRCA committee member
Carolina Roofing and Sheet Metal Contractors Association president
Spouse: Jane; children: Melinda, Pam, Susan.

1962

CHARLES N. GRIFFITHS SR. (1915–1986)
Binghamton Slag Roofing Co., Inc., Binghamton, N.Y.
NRCA committee member
NRCA J. A. Piper Award 1982
Master's degree in science, Massachusetts Institute of Technology; professional engineer, registered in New York
Spouse: Elizabeth; children: Charles Jr., Sally, Libby, Peter.

1963

MILTON J. OLSON (1910–)
Olson Brothers, Inc., Omaha, Neb.
NRCA Committee on Technical Services and Research; NRCA liaison to American Society for Testing and Materials
NRCA J. A. Piper Award 1966
Midwest Roofing Contractors Association president 1958; James Q. McCawley Award 1975; Omaha Sheet Metal Contractors Association president
Spouse: Ann; children: Donald, Sandra.

1964

JOHN E. HARTMANN (1913–1976)
Hartmann Company, Inc., Terre Haute, Ind.
NRCA committee member; Foremen and Superintendents Conference chair 1959
NRCA J. A. Piper Award 1968
Indiana Contractors Association president 1960
Spouse: Thelma; children: Joan, Ann, John, Nanci, James.

1965

GEORGE K. BURRUS (1918–)
Tri-State Roofing Company of Tennessee, Knoxville, Tenn.
NRCA committee chair
NRCA J. A. Piper Award 1971
Tennessee Association of Roofing Contractors first president; Knoxville Roofing and Sheet Metal Association president, secretary and treasurer
Spouse: Dorothy; children: George Jr., Sarah, Judy.

1966

ROBERT E. BUBENZER (1912–)
Henry C. Smither Roofing Co., Inc., Indianapolis, Ind.
NRCA committees on technical services and research and manual
NRCA J. A. Piper Award 1964
Indiana Subcontractor Association first president; Indiana Roofing Contractor Association president
Spouse: Mary Eloise; children: Robert, Bruce, Brenda.

1967

J. J. HALL JR. (1909–1977)
Sechrist-Hall Co., Corpus Christi, Texas
NRCA committee member
NRCA J. A. Piper Award 1970
Spouse: Maurine; children: Judith, Pamela, J. J. III.

1968 .... **EDWARD D. WEYAND** Sacramento, Calif.

1969

HENRY E. ALCOCK (1911–)
M. W. Powell Co., Chicago, Ill.
NRCA committee member
Spouse: Marion; children: Mike, Judy.

1970

BENNETT HUTCHISON JR. (1924–)
Tip Top Roofers, Inc., Atlanta, Ga.
NRCA committee chair
Master's degree in science
Spouse: Myrtice; children: Bennett III, Clint, Mark, Jace.

1971

ROBERT E. OSTERHOLT (1920–1974)
South Side Roofing and Sheet Metal Co., Inc., St. Louis, Mo.
NRCA committees on labor, budget and finance and insurance
NRCA J. A. Piper Award 1975 (posthumous)
Midwest Roofing Contractors Association vice president, secretary-treasurer; MRCA James Q. McCawley Award 1974; Roofing Contractors of Greater St. Louis president 1963–1965
Spouse: Henrietta; children: Carolyn, Robert, William.

1972–1973

JAMES F. KING (1928–)
Snyder Roofing and Sheet Metal Inc., Portland, Ore.
NRCA Dues Review Committee chair; Safety Committee chair; Membership Committee; Subcommittee to Technical Services and Research
Portland Contractors Association president
Children: Scott, Holly, Kyle.

1973

WILLIAM R. STEINMETZ (1926–)
Midland Engineering Co., Inc., South Bend, Ind.
NRCA Membership Committee
NRCA J. A. Piper Award 1976
Midwest Roofing Contractors Association president; Indiana Roofing Contractors Association president; MRCA James Q. McCawley Award; Roofing Industry Education Institute Board of Regents chair
Spouse: Shirley; children: Janet, William Jr.

1974

CHARLES J. RAYMOND (1922–)
Giffen Roofing Co., Miami, Fla.
National Roofing Foundation chair; National Roofing Legal Resource Center president 1980–81
NRCA J. A. Piper Award 1977
Namesake of Charlie Raymond membership award 1980
Florida Roofing and Sheet Metal Association president; FRSA Bob Campanella Award
Spouse: Ruth; children: Lynda, Charlyne.

1975

ROBERT E. LINCK (1931–)
Robert E. Linck, Inc., Philadelphia, Pa.
NRCA committee member
NRCA J. A. Piper Award 1978
Roofing Craft Board chair under the CISC for the stabilization of wages and prices in the construction industry
Life fellow, Royal Society for the Encouragement of Arts, Manufactures and Commerce, London
Spouse: Annette; children: Vanessa, Robert.

1976

GEORGE E. STEPHENSON (1921–)
Stephenson Roofing Co., St. Louis, Mo.
NRCA committee member
NRCA J. A. Piper Award 1985
Midwest Roofing Contractors Association president 1972; Roofing Contractors of Greater St. Louis president
Spouse: Lila; children: George Jr., Brian, Thomas, James, John.

1977

J. ROY MARTIN JR. (1918–)
J. Roy Martin & Co., Inc., Anderson, S.C.
NRCA first senior vice president; committee member; Second International Symposium presiding officer
NRCA J. A. Piper Award 1986
Carolinas Roofing and Sheet Metal Association president; CRSMA Gordon M. Waters Distinguished Service Award
Spouse: Alma; children: J. Roy III, John, Stephen, Margaret.

1978

MELVIN I. KRUGER (1929–)
L. E. Schwartz & Son, Inc., Macon, Ga.
NRCA committee member; Roofing Systems Technical Committee chair 1980–82; American Institute of Architects liaison chair 1981–83; Associated General Contractors liaison chair 1981–83; Thermal Insulation Manufacturers Association liaison chair 1980–83
NRCA J. A. Piper Award 1980
Roofing and Sheet Metal Contractors of Georgia president 1970–71
Roofing Industry Education Institute Board of Regents chair 1979–80
Spouse: Beverly; children: Lynn, Steven, Gail.

1979

THOMAS G. MANSON (1926–)
Western Roofing Co., Kansas City, Mo.
NRCA committee member; Roofing Systems Technical Committee co-founder and first chair; National Roofing Foundation trustee
Midwest Roofing Contractors Association president
Spouse: Frances; children: Jenny, Thomas, Marcus, Theodore.

1980

WILLIAM E. KUGLER (1935–)
United Roofing and Waterproofing, Inc., Denver, Colo.
NRCA Roofing and Waterproofing Manual Committee chair 1979–80; Technical and Research Committee chair 1979–80; Roofing Systems Technical Committee chair 1978–80
NRCA J. A. Piper Award 1979
Midwest Roofing Contractors Association president 1969; James Q. McCawley Award 1970
Roofing Industry Education Institute faculty member 1983–86
Spouse: Zoe; child: Lisa.

1981

JOHN ZAMRZLA (1939–)
Western Pacific Roofing Corp., Lancaster, Calif.
NRCA Government Relations Committee chair; Air Pollution Committee chair; Safety Committee chair; Long Range Planning Committee
NRCA J. A. Piper Award 1984
Roofing Contractors of Southern California president; Western States Roofing Contractors Association president; Roofing Contractors Association of California president
Spouse: Pam; children: John, Joey, Sheri.

1982

JOHN W. BRADFORD (1929–)
Bradford Roofing and Insulation, Billings, Mont.
NRCA Technical Services and Research Committee chair; Budget and Finance Committee
Midwest Roofing Contractors Association president 1976; MRCA James Q. McCawley Award
Roofing Industry Education Institute Board of Regents vice chair
Spouse: LaVette; children: Tracy, Dane, Thomas.

1983

BURTON J. KARP (1936–)
The Eagle Group, West Hartford, Conn.
NRCA Technical Operating Committee; Foremen and Superintendents Conference chair
Northeast Roofing Contractors president
Roofing Industry Education Institute vice chair
University of Hartford-Construction Institute chair
Children: Jamie, Dale, Michael.

1984

WAYNE I. MULLIS (1940–)
Universal Roofers Inc., Phoenix, Ariz.
NRCA committees on budget and finance, membership, education, long-range planning
NRCA Charlie Raymond Award
Arizona Roofing Contractors Association president; Western States Roofing Contractors Association president
Spouse: Janie; children: Stephen, Christie, Jason, Christina.

1985

ROBERT T. HARRISON (1922–)
Greenville Roofing Co., Inc., Greenville, S.C.
NRCA committees on budget and finance, education, public relations, performance evaluation, convention, Foremen and Superintendents Conference, BUR task force; speakers bureau
Carolinas Roofing Association president 1973–1974
Spouse: Olive; children: Janet, Ann, Linwood.

THE J. A. PIPER AWARD WINNERS

The J. A. Piper Award, the highest honor annually awarded in the American roofing industry, is "in recognition of devoted and constant outstanding service."

J. A. Piper, NRCA president in 1932–33, weathered exceptional financial circumstances during the nation's Depression. It was his responsibility to hold the Association and its members together, and he rose to the occasion. In appreciation for his legacy of dedication to NRCA, his industry, and his country, the Piper family of Greenville, South Carolina established the memorial award in 1947.

1948	Irwin Langer
1950	Tom Daly
1951	M. T. Buckley
1954	J. Boyd Griffiths
1955	Hilton Bowles
1956	Joe Mattingly
1957	Gordon Waters
1958	Clarence Esbenshade
1959	Joseph Kovarsky
1960	Walter Simon
1961	Clyde Scott
1962	John Reuter
1963	Calvin Bowman
1964	R. E. Bubenzer
1965	J. W. Kerr
1966	Milton Olson
1967	Clarence Carr
1968	John Hartmann
1969	L. A. Kautz
1970	J. J. Hall
1971	George Burrus
1972	George Bodwell
1973	Paul Morris
1974	William Cullen
1975	Robert Osterholt
1976	William Steinmetz, Sr.
1977	Charles Raymond
1978	Robert Linck
1979	William Kugler
1980	Melvin Kruger
1981	Fred Good
1982	Charles Griffiths, Sr.
1983	Sam Piper
1984	John Zamrzla
1985	George Stephenson
1986	J. Roy Martin, Jr.

Those individuals or groups not intimately familiar with NRCA are always amazed to learn that Association members have been in business an *average* of 35 years. In an industry too often characterized by slap-dash start-ups and hasty departures, these businessmen and businesswomen stand as a testament to reliability and skilled workmanship. Those companies who have more recently joined the ranks of professional roofing contractors may not be able to claim this lengthy record of service, but their membership in NRCA demonstrates the value they place on these same qualities. Whether you are celebrating time-honored traditions in your firm or just hanging out your shingle, the National Roofing Contractors Association salutes you.

The following list is composed of companies that responded to questionnaires or telephone inquiries originating from Association headquarters from 1983 through 1986. Companies are listed alphabetically within chronological order.

1795–1887

	Joined NRCA	No. of Generations	Founded
Bird Inc. Walpole, MA	1972	–	1795
Du Pont Co. Wilmington, DE	1980	–	1802
City of Norfolk Norfolk, VA	1985	–	1845
M. W. Powell Co. Chicago, IL	1953	6	1847
Nicholson and Galloway Inc. Glen Head, NY	1969	4	1849
The Wehner Roofing and Tinning Co. Dayton, OH	1955	5	1850
Warren-Ehret Co. of Maryland Inc. Blue Bell, PA	1951	–	1852
W. R. Grace and Co. Cambridge, MA	1970	–	1854
Michigan State University East Lansing, MI	1984	–	1855
Abbott-Sommer Inc. Bronx, NY	1969	–	1857
ATEC Associates Inc. Indianapolis, IN	1983	–	1858
Fred S. James and Co. Chicago, IL	1976	–	1858
Manville Roofing Systems Division Denver, CO	1967	–	1858
Vermont Structural Slate Co., Inc. Fair Haven, VT	1977	–	1859
M. B. Bourne and Son Inc. South Portland, ME	1970	2	1860
The New York Roofing Co. Woodside, NY	1953	–	1860

	Joined NRCA	No. of Generations	Founded
Roofing Systems Inc. Pawtucket, RI	1976	–	1864
Bruno Martin Co. Saginaw, MI	1972	2	1865
Milliken and Co. La Grange, GA	1984	–	1865
Isaacson and Sons Inc. Des Moines, IA	1981	5	1866
Merchant and Evans Industries Inc. Burlington, NJ	1979	–	1866
Max C. Smith Co. Gibsonia, PA	1978	3	1868
Henry C. Smither Roofing Co., Inc. Indianapolis, IN	1953	3	1868
Rising and Nelson Slate Co., Inc. West Pawlet, VT	1963	4	1869
Grove Roofing Co., Inc. Buffalo, NY	1971	4	1870
N. V. Lummerzheim and Co. Belgium	1985	–	1870
Seaman and Schuske Metal Works Co. St. Joseph, MO	1983	5	1870
Central City Roofing Co., Inc. Syracuse, NY	1967	–	1872
Henry R. Fell Co. Trenton, NJ	1979	–	1872
Winding Roofing Co., Inc. Milwaukee, WI	1953	2	1873

	Joined NRCA	No. of Generations	Founded
Drake Roofing Inc. Des Moines, IA	1985	3	1875
A. C. Hathorne Co., Inc. Williston, VT	1964	–	1875
H. C. Truitt Co. Covington, KY	1966	–	1875
Knickerbocker Roofing/Paving Co. Harvey, IL	1953	4	1877
The Austin Co. Cleveland, OH	1984	–	1878
Sellers and Marquis Roofing Co. Kansas City, MO	1953	2	1878
J. D. Candler Roofing Co., Inc. Detroit, MI	1981	–	1879
F. J. A. Christiansen Roofing Co., Inc. Milwaukee, WI	1925	–	1879
The Imbus Roofing Co., Inc. Cold Spring, KY	1953	5	1879
Watchtower Bible and Tract Society Brooklyn, NY	1985	–	1879
F. J. Dahill Co., Inc. New Haven, CT	1968	4	1883
Curran Roofing Co. Marion, IN	1946	–	1884
Ketcher and Co., Inc. North Little Rock, AR	1985	5	1885
Palm Beach County Schools West Palm Beach, FL	1986	–	1885
Consumers Roofing Co., Inc. Hammond, IN	1964	5	1886

	Joined NRCA	No. of Generations	Founded
William H. Lavey and Associates El Toro, CA	1982	6	1886
Allendorfer Roofing Co., Ltd. Chicago, IL	1981	2	1887
Dunne Roofing Co. Chicago, IL	1976	3	1887
Ludowici-Celadon Co. New Lexington, OH	1981	–	1887

1888–1913

	Joined NRCA	No. of Generations	Founded
Abram Roofing Co. Louisville, KY	1967	5	1888
Carlson Roofing Co., Inc. Rockford, IL	1953	4	1888
The G. R. Cummings Co. Meriden, CT	1955	4	1888
Detroit Cornice and Slate Co. Ferndale, MI	1977	4	1888
Dove Roofing Inc. Tallahassee, FL	1974	3	1888
Jamestown Roofing Inc. Jamestown, NY	1983	2	1888
The Charles E. Mahaney Roofing Co. Wichita, KS	1981	–	1888
The Norton Brothers Co. Cleveland, OH	1960	3	1888
E.W. Roberts Sons Inc. Luzerne, PA	1974	3	1888
Frank Staar and Sons Inc. Glenview, IL	1953	3	1888
Beidler-Taylor Roofing Co. Ashtabula, OH	1969	3	1889
Henry Bireline Co., Inc. Danville, IL	1974	4	1889
James Mansfield and Sons Co., Inc. Lyons, IL	1953	3	1890
Peerless Roofing Co., Ltd. Honolulu, HI	1983	–	1890
George Rogerson Ltd. England	1986	4	1890
F. O. Schoedinger Inc. Columbus, OH	1964	2	1890
Shive-Hattery Engineers Inc. Cedar Rapids, IA	1984	–	1890
American Roofing and Repair Co. West Chicago, IL	1981	–	1891
Lydick Hooks Roofing Co., Inc. Abilene, TX	1973	3	1891
Lydick Roofing Co. of Brownwood TX Inc. Brownwood, TX	1977	–	1891
Lydick Roofing Co. Midland, TX	1979	–	1891
Kahlmeyer Brothers Inc. St. Louis, MO	1983	4	1892
The Enterprise Cos. Wheeling, IL	1983	3	1893
C. L. Linfoot Co. Grand Forks, ND	1971	4	1893
South Side Roofing Corp. Milwaukee, WI	1953	4	1893

	Joined NRCA	No. of Generations	Founded
Western Roofing Co. Kansas City, MO	1970	–	1893
Champion International Hamilton, OH	1985	–	1895
The Garland Co. Cleveland, OH	1976	–	1895
King Roofing and Manufacturing Co. Sanford, NC	1966	3	1895
Luigi Menestrina Trento, Italy	1986	–	1895
Young Sales Corp. St. Louis, MO	1958	3	1895
Atlas Bolt and Screw Co. Bensenville, IL	1983	–	1896
Ingold Co., Inc. Hickory, NC	1953	3	1896
Chamberlin Waterproofing Inc. Kansas City, MO	1984	1	1897
Chamberlin Waterproofing and Roofing Systems Houston, TX	1984	–	1897
Ralph R. Reeder and Sons Inc. Indianapolis, IN	1953	–	1897
Peter W. Eberz and Son Inc. Buffalo, NY	1978	3	1898
Nelson Roofing Co., Inc. Chicago, IL	1976	4	1898
Olson Brothers Inc. Omaha, NE	1953	3	1898

	Joined NRCA	No. of Generations	Founded
Seline Sheet Metal Works Inc. Houston, TX	1972	3	1898
Standard Roofing Co., Inc. Oklahoma City, OK	1953	2	1898
Becker Roofing Co. Madison, WI	1981	–	1899
American Associated Co. Atlanta, GA	1969	–	1900
Baschnagel Brothers Inc. Whitestone, NY	1985	3	1900
Karl H. Frye Inc. Bluefield, WV	1980	2	1900
Fred Hartel Co., Inc. Galveston, TX	1971	–	1900
George Mehrer and Son Inc. Conshohocken, PA	1974	3	1900
Mooi Roofing Co., Inc. Holland, MI	1981	4	1900
Pulver Roofing Co., Inc. Utica, NY	1973	3	1900
Reichel and Drews Inc. Itasca, IL	1985	–	1900
Wharton Roofing Co. Parkersburg, WV	1985	2	1900
Eagle Cornice Co., Inc. Cranston, RI	1968	3	1901
Hollander and Co., Inc. Highland Park, IL	1971	–	1901
Easton Roofing Co., Inc. Easton, PA	1979	3	1902

	Joined NRCA	No. of Generations	Founded
Fisher Scientific Co. Pittsburgh, PA	1984	–	1902
H. Klein and Sons Inc. Mineola, NY	1977	–	1902
Sutter Roofing and Metal Co., Inc. Clarksburg, WV	1973	3	1902
Winona Heating and Ventilating Co. Winona, MN	1974	3	1902
The Zero-Breese Co. Cincinnati, OH	1954	2	1902
The Bellet Roofing Co., Inc. Bronx, NY	1986	3	1903
Hanover Iron Works Inc. Wilmington, NC	1974	–	1903
Hoekstra Roofing Co. Kalamazoo, MI	1983	3	1903
S. Kane and Son Inc. Philadelphia, PA	1967	–	1903
A. J. Shirk Roofing Co. Kansas City, MO	1974	3	1903
American Roofing and Metal Co., Inc. San Antonio, TX	1978	4	1904
AB Mataki Malmo, Sweden	1986	–	1904
Pickens Roofing and Sheet Metal Inc. Spartanburg, SC	1979	–	1904
G. G. Ray Co. Charlotte, NC	1953	2	1904

	Joined NRCA	No. of Generations	Founded
Shelton Roofing Co., Inc. Derby, CT	1953	–	1904
Aronow Roofing Co. Camden, NJ	1967	3	1905
Del/Mac Roofing Inc. Columbia, SC	1953	–	1905
Kimmenade International B.V. Netherlands	1969	3	1905
Steyer Roofing Co. Warren, MI	1964	4	1905
Jim Taylor Inc. Belleville, IL	1967	2	1905
A. Wachsberger Roofing and Sheet Metal Works Inc. Lynbrook, NY	1977	3	1905
Chris Andersen Roofing Perth Amboy, NJ	1983	4	1906
Associated Dry Goods Corp. New York, NY	1980	–	1906
Sherriff-Goslin Co. Battle Creek, MI	1972	3	1906
Koppers Co., Inc. Pittsburgh, PA	1965	–	1907
The Lawson Roofing Co., Inc. San Francisco, CA	1976	3	1907
M. M. Schranz Roofing Inc. Milwaukee, WI	1953	–	1907
Schriber Roofing Co. Dayton, OH	1966	3	1907
J. E. Wood and Sons Co., Inc. Clinton, MD	1974	4	1907

	Joined NRCA	No. of Generations	Founded
Morris Black and Sons Inc. Bethlehem, PA	1986	4	1908
J. W. Brooks and Sons Chattanooga, TN	1953	3	1908
H. T. Harrison and Sons Inc. Rockville, MD	1984	–	1908
J. A. Piper Roofing Co. Greenville, SC	1953	3	1908
Railton Manufacturing Co. Houston, TX	1971	2	1908
Babcock-Davis Hatchways Inc. Arlington, MA	1983	2	1909
Homasote Co. West Trenton, NJ	1986	–	1909
Permanite Ltd. England	1982	–	1909
Tuscher Roofing Co. Oak Park, IL	1953	3	1909
A-1 Roofing Co. Chicago, IL	1982	3	1910
Samuel Dean Co. San Antonio, TX	1972	–	1910
Gupton Sheet Metal South Boston, VA	1983	3	1910
Luppold Roofing Co., Inc. Reading, PA	1968	–	1910
J. D. Miles and Sons Inc. Chesapeake, VA	1964	–	1910
L. E. Schwartz and Son Inc. Macon, GA	1964	4	1910

	Joined NRCA	No. of Generations	Founded
ATAB N.V. Belgium	1985	–	1911
Dessent Roofing Co., Inc. Chicago, IL	1966	3	1911
Passaic Metal Products Co. Clifton, NJ	1965	4	1911
Ruberoid Building Products England	1978	–	1911
Solomon and Son Sheet Metal Co. Flint, MI	1981	3	1911
Chicago Testing Laboratory Inc. Northbrook, IL	1983	–	1912
John Connelly Roofing Co. Valley Park, MO	1975	3	1912
John Dalsin and Son Inc. Minneapolis, MN	1958	–	1912
The Dansky Corp. West Hartford, CT	1968	3	1912
N. L. Freedman Inc. Springfield, MA	1981	–	1912
R. M. Lucas Co. Chicago, IL	1983	–	1912
R. A. Lysy Roofing Co. Phoenix, AZ	1984	2	1912
Malott and Peterson Roofing Co. Berkeley, CA	1979	3	1912
Marathon Roofing Products Buffalo, NY	1973	–	1912
National Varnish Co. Detroit, MI	1985	3	1912

	Joined NRCA	No. of Generations	Founded
Wm. J. Schmitt Inc. Rochester, NY	1980	3	1912
T. F. Sloan Co. Paterson, NJ	1953	–	1912
Stockton Roofing Co. Stockton, CA	1953	3	1912
Universal General Sheet Metal Inc. Edison, NJ	1963	–	1912
Fred K. Wallace and Son Inc. Philadelphia, PA	1973	2	1912
M. Walter and Co. Chicago, IL	1980	3	1912
Binghamton Slag Roofing Co., Inc. Binghamton, NY	1954	4	1913
Garey Roof Inc. St. Joseph, MI	1969	4	1913
The Gibson Homans Co. Twinsburg, OH	1970	–	1913
S. M. Kisner and Sons Fairmont, WV	1953	–	1913
L. Marshall Inc. Glenview, IL	1982	–	1913
N. W. Martin and Brothers Inc. Richmond, VA	1953	–	1913
Salina Roofing Inc. Salina, KS	1967	–	1913

1914–1939

	Joined NRCA	No. of Generations	Founded
Avon Sheet Metal Roofing Co. Newark, NJ	1980	3	1914
Budd Piper Roofing Co. Durham, NC	1953	3	1914
Knapp Roofing Co., Inc. Three Rivers, MI	1981	1	1914
Olsson Roofing Co., Inc. Aurora, IL	1971	2	1914
Turner Roofing Co. Crossville, TN	1982	–	1914
Turner Roofing Co. San Antonio, TX	1953	–	1914
J. S. Wagner Co., Inc. Hyattsville, MD	1972	–	1914
Augusta Roofing and Metal Works Inc. Augusta, GA	1961	–	1915
Baker Roofing Co. Raleigh, NC	1953	3	1915
Blackwell Burner Co. San Antonio, TX	1963	–	1915
Nuroy Roofing and Sheet Metal Inc. Orrville, OH	1975	3	1915
Southwestern Sheet Metal Works Inc. El Paso, TX	1983	1	1915
Vancouver Roofing and Sheet Metal Co. Vancouver, WA	1973	–	1915
Western Waterproofing Co., Inc. St. Louis, MO	1986	–	1915
Alta Roofing Co. San Francisco, CA	1968	3	1916

	Joined NRCA	No. of Generations	Founded
Atlas Roofing Corp. Houston, TX	1967	–	1916
Evergreen Slate Co., Inc. Granville, NY	1970	3	1916
Hahnel Brothers Co. Lewiston, ME	1959	3	1916
Hartmann Co., Inc. Terre Haute, IN	1953	3	1916
McKinley Roofing Inc. Mansfield, OH	1967	–	1916
Louis T. Ollesheimer and Son Inc. Madison Heights, MI	1965	3	1916
Owen Roofing Co. Los Angeles, CA	1968	3	1916
Pancratz Co. Casper, WY	1960	–	1916
M. I. Sachs and Son Roofing Inc. Baltimore, MD	1973	2	1916
Schreiber Corp. Detroit, MI	1953	–	1916
Thompson Roofing and Supply Co. Enid, OK	1985	4	1916
Wrisco Industries Inc. Linden, NJ	1977	–	1916
Aeroil Products Co., Inc. South Hackensack, NJ	1963	–	1917
Carlisle SynTec Systems Carlisle, PA	1974	–	1917
Meyers Roofing Service Yuba City, CA	1979	3	1917

	Joined NRCA	No. of Generations	Founded
Parker Roofing Co. San Antonio, TX	1980	–	1917
J. P. Patti Co., Inc. Saddle Brook, NJ	1963	4	1917
John F. Shea Co., Inc. Mattapan, MA	1963	3	1917
George Shustick and Sons Inc. Columbus, OH	1964	–	1917
Long Island Tinsmith Supply Corp. Richmond Hill, NY	1965	3	1918
L. Martone and Sons Inc. Glen Cove, NY	1971	–	1918
Newton and Crane Inc. Pontiac, MI	1967	–	1918
Tuschyn Roofing Co., Inc. Roselle, NJ	1979	3	1918
John Cole and Sons Inc. Baltimore, MD	1968	3	1919
Deerland Corp. Norristown, PA	1986	–	1919
The Eagle Group Inc. West Hartford, CT	1983	2	1919
Eagle Moisture Protection Corp. West Hartford, CT	1964	2	1919
Guarantee Roofing Co., Inc. Terre Haute, IN	1953	4	1919
Haight Roofing Co., Inc. Seattle, WA	1979	4	1919
Independent Roofing Co., Inc. Omaha, NE	1972	1	1919

	Joined NRCA	No. of Generations	Founded
Jos. A. Sanders and Sons Inc. Buffalo, NY	1979	–	1919
Sanders Roofing Co., Inc. Union, NJ	1964	3	1919
Stroup Sheet Metal Works Inc. Ashville, NC	1964	3	1919
All-Weather Roofing Co., Inc. Buffalo, NY	1968	3	1920
Julien P. Benjamin Equipment Co. Jacksonville, FL	1973	–	1920
R. T. Brundage Roofing Co., Inc. Kalamazoo, MI	1974	–	1920
Dubois Building Products Cincinnati, OH	1979	–	1920
J. B. Eurell Co. Edgemere, MD	1974	2	1920
Falkner Inc. Orlando, FL	1972	2	1920
Hoge-Warren-Zimmermann of Detroit Detroit, MI	1974	3	1920
Hoge-Warren-Zimmermann Co. Rolling Meadows, IL	1974	3	1920
Indiana Bell Telephone Co., Inc. Indianapolis, IN	1983	3	1920
Jones and Cleary Roofing Co., Inc. Chicago, IL	1982	3	1920
Kirberg Roofing St. Louis, MO	1975	3	1920
Langer Roofing and Sheet Metal Inc. Milwaukee, WI	1956	–	1920

	Joined NRCA	No. of Generations	Founded
Leamon Roofing and Sheet Metal Inc. Center Valley, PA	1984	–	1920
Mastercraft Tile and Roofing Co. Richmond, CA	1980	–	1920
Morris Roofing and Sheet Metal Corp. Springfield, MA	1965	–	1920
Reader Tinning Roofing and Heating Co. Cleveland, OH	1985	3	1920
Ross and Barrows Inc. Santa Ana, CA	1983	2	1920
B. Sheber and Sons Inc. Albany, NY	1967	3	1920
South Side Roofing Co., Inc. St. Louis, MO	1959	3	1920
Thomas Roofing and Sheet Metal Co., Inc. Atlantic City, NJ	1979	3	1920
Wooster Sheet Metal and Roofing Co. Akron, OH	1953	4	1920
A-1 Roofing Service Inc. Los Angeles, CA	1974	2	1921
American Sheet Metal Corp. Norfolk, VA	1953	3	1921
E. D. Chase Co., Inc. Oshkosh, WI	1953	–	1921
Dealers Supply Co. Portland, OR	1978	3	1921

	Joined NRCA	No. of Generations	Founded
Dependon Exteriors Inc. Chicago, IL	1985	3	1921
Gate City Roofing Co., Inc. Greensboro, NC	1953	–	1921
Goslee Roofing and Sheet Metal Inc. Salisbury, MD	1968	2	1921
Horel-George Co. Eau Claire, WI	1970	1	1921
Lafayette Roofing and Sheet Metal Lafayette, IN	1974	–	1921
Lone Star Systems Inc. Austin, TX	1985	–	1921
M. J. Murphy and Sons Inc. Dover, NH	1980	–	1921
C. C. Olipant and Son Inc. Laurel, DE	1975	–	1921
Palisade Roofing Co. Chicago, IL	1958	3	1921
Park Roofing Co., Inc. New Brunswick, NJ	1963	3	1921
Pellow Roofing and Sales Inc. Marquette, MI	1981	–	1921
The Quality Roofing Co. Kansas City, MO	1972	3	1921
Rawl Co., Inc. New Rochelle, NY	1984	3	1921
Standard Roofing Co. Montgomery, AL	1960	3	1921
L. D. Sterns Corp. Cleveland, OH	1984	3	1921

	Joined NRCA	No. of Generations	Founded
Acme Roofing and Sheet Metal Cleveland, OH	1986	–	1922
Elco Industries Inc. Const. Prod. Rockford, IL	1983	–	1922
Frank P. Frey and Co. Melrose Park, IL	1979	3	1922
Edward R. Hart Co. Canton, OH	1969	–	1922
J-Co Equipment Bridgeview, IL	1985	–	1922
Kreiling Roofing Co. Peoria, IL	1984	3	1922
Leroy Roofing Co. Harrisburg, PA	1953	–	1922
Mid-Valley Roofing and Supply Co., Inc. Moosic, PA	1953	2	1922
Midland Engineering Co., Inc. South Bend, IN	1953	3	1922
Missouri Roofing Co. St. Louis, MO	1953	2	1922
R and J Simpson Roofing Co., Inc. Paterson, NJ	1978	3	1922
Rainbow Roofing Co. Chicago, IL	1958	3	1922
C. L. Schust Co., Inc. Fort Wayne, IN	1963	–	1922
Snyder Roofing and Sheet Metal Inc. Tigard, OR	1959	–	1922
Texas Refinery Corp. Fort Worth, TX	1983	3	1922

	Joined NRCA	No. of Generations	Founded
Warren Roofing and Insulating Co. Walton Hills, OH	1954	–	1922
AAA Roofing Co., Inc. Indianapolis, IN	1978	3	1923
Boys Roofing and Air Conditioning Inc. West Palm Beach, FL	1969	–	1923
Guaranteed Roofing and Sheet Metal Co. Cleveland, OH	1953	3	1923
Karnak Chemical Corp. Clark, NJ	1979	3	1923
Kaw Roofing and Sheet Metal Inc. Kansas City, KS	1953	3	1923
John H. Kenney Roofing and Sheet Metal Co. Dover, NJ	1974	2	1923
Malcor Roofing Co. Aurora, IL	1973	–	1923
Range Cornice and Roofing Co. Hibbing, MN	1974	3	1923
Roth Brothers Inc. Youngstown, OH	1967	–	1923
Sobel and Kraus Inc. Bronx, NY	1969	3	1923
Tri-State Roofing and Sheet Metal Co. Lexington, KY	1968	3	1923
Tri-State Roofing and Sheet Metal Co. Cumberland, MD	1958	3	1923

	Joined NRCA	No. of Generations	Founded
Tri-State Roofing and Sheet Metal Co. Davisville, WV	1952	3	1923
Tri-State Roofing and Sheet Metal Co. Charleston, WV	1953	3	1923
United Roofing and Waterproofing Inc. Denver, CO	1968	3	1923
Virgin Roof Co. San Gabriel, CA	1970	–	1923
Allied Roofers Supply Corp. East Rutherford, NJ	1979	–	1924
Allied Roofing and Siding Co. Grand Rapids, MI	1975	–	1924
Bradley and Sons Inc. Lancaster, NY	1968	3	1924
Celanese Industrial Fibers Charlotte, NC	1983	–	1924
Covington Roofing Co., Inc. Conway, AR	1984	–	1924
Cronin Asphalt Corp. East Providence, RI	1979	3	1924
Florida Sheet Metal Inc. Lakeland, FL	1979	–	1924
The Hartford Roofing Co., Inc. Glastonbury, CT	1972	3	1924
B. Hoffman Roofers Inc. Buffalo, NY	1967	–	1924
Izmirian Roofing and Sheet Metal San Mateo, CA	1974	–	1924

	Joined NRCA	No. of Generations	Founded
Scally Waterproofing Co. St. Louis, MO	1964	–	1924
Slatile Roofing and Sheet Metal Co., Inc. South Bend, IN	1973	4	1924
Star Roofing Co., Inc. Oakland, CA	1980	2	1924
Albany Sheet Metal Works Albany, GA	1982	–	1925
American Pacific Roofing San Diego, CA	1981	–	1925
Hageman Roofing Co. Ridgefield Park, NJ	1976	2	1925
Linenthal Eisenberg Anderson Inc. Boston, MA	1984	–	1925
Lydick Roofing Co. of Lubbock Inc. Lubbock, TX	1975	3	1925
MacLean Fogg Co. Richmond, IL	1984	3	1925
Yates Roofing Corp. New York, NY	1975	3	1925
J. L. Adler Roofing Inc. Joliet, IL	1953	–	1926
Bash Pepper Roofing Co. Champaign, IL	1983	1	1926
C. O. Beck and Sons Inc. Waterville, ME	1970	3	1926
Biebel Brothers Inc. Roofing Contractors St. Louis, MO	1953	–	1926

	Joined NRCA	No. of Generations	Founded
Boice Roofing Co. Westmont, IL	1953	3	1926
Brisk Waterproofing Co., Inc. Ridgefield, NJ	1979	–	1926
City Roofing Co. Elkhart, IN	1953	3	1926
Enterprise Roofing and Sheet Metal Co. Dayton, OH	1953	3	1926
Georgia Pacific Corp. Atlanta, GA	1982	–	1926
Georgia Power Co. Atlanta, GA	1985	–	1926
Graco Inc. Minneapolis, MN	1980	–	1926
Keystone Roofing Co., Inc. Pennsauken, NJ	1958	3	1926
Parichy Roofing and Shingle Co. Oak Park, IL	1958	3	1926
Pittman-Waller Roofing Co., Inc. Macon, GA	1953	–	1926
Wm. Schaus and Son Inc. Manitowoc, WI	1983	4	1926
Southern California Roofing Co. Downey, CA	1971	–	1926
Stiles Roofing Inc. Lebanon, MO	1979	–	1926
Babsons Inc. Orwigsburg, PA	1979	3	1927
Bacon and Schramm Inc. Denver, CO	1960	2	1927

	Joined NRCA	No. of Generations	Founded
Joseph T. Cazeault and Sons East Weymouth, MA	1964	4	1927
Economy Roofing Co. Bettendorf, IA	1967	–	1927
Haning's Inc. Lancaster, OH	1983	3	1927
Henderson-Johnson Co., Inc. Syracuse, NY	1974	3	1927
Nordheim Roofing and Sheet Metal Co. Bemidji, MN	1978	–	1927
Roofing and Supply Co. Chattanooga, TN	1979	3	1927
Simmons Roofing Co., Inc. Grand Rapids, MI	1953	3	1927
Stewart-Barry Roofing and Insulation Co. South Holland, IL	1976	–	1927
Tri-State Roofing Inc. Scottsbluff, NE	1981	2	1927
Twin City Roofing and Material of Mandan Mandan, ND	1986	3	1927
Twin City Roofing of Wahpeton Inc. Wahpeton, ND	1972	3	1927
Bend Industries Inc. West Bend, WI	1985	–	1928
Cheney Flashing Co. Trenton, NJ	1985	3	1928
Dahm Brothers Inc. Fort Wayne, IN	1972	3	1928

	Joined NRCA	No. of Generations	Founded
J. Roy Martin and Co., Inc. Anderson, SC	1953	3	1928
Mont Clare Roofers Inc. Chicago, IL	1981	–	1928
Orleans Sheet Metal and Roofing New Orleans, LA	1983	5	1928
Poole Roofing and Sheet Metal Co. Gainesville, FL	1953	3	1928
Reger Roofing and Siding Co. Kirkwood, MO	1973	–	1928
Roberts Roof and Floor Inc. Las Vegas, NV	1972	2	1928
Southern Blow Pipe and Roofing Co. Chattanooga, TN	1967	–	1928
The Stolle Corp. Sidney, OH	1986	–	1928
Tremco Inc. Cleveland, OH	1971	–	1928
U.S. Borax and Chemical Corp. Boron, CA	1984	–	1928
Young Roofing Co., Inc. Durham, NC	1981	–	1928
Acme Roofing Co., Inc. Auburn, MA	1954	3	1929
Bangor Roofing and Sheet Metal Co. Bangor, ME	1968	–	1929
Buckingham-Virginia Slate Corp. Richmond, VA	1963	4	1929
BASF Corp. Fibers Division Enka, NC	1985	–	1929

	Joined NRCA	No. of Generations	Founded
Central Roofing Co. Minneapolis, MN	1960	3	1929
Consumers Roofing and Insulation Works Chicago, IL	1953	4	1929
Frederic Roofing and Sheet Metal Co., Inc. St. Louis, MO	1953	3	1929
Geissler Roofing Co., Inc. Belleville, IL	1953	2	1929
Gleason Roofing Inc. Philadelphia, PA	1984	–	1929
Home Roofing Tampa, FL	1981	2	1929
Nassau Roofing and Sheet Metal Co., Inc. Garden City Park, NY	1981	3	1929
Picquet's Roofing Inc. Charleston, SC	1982	2	1929
Trinity Engineering Testing Corp. Ft. Hood, TX	1985	–	1929
Contractors Group Augusta, ME	1979	–	1930
Daniel Construction Co. Greenville, SC	1973	–	1930
Independent Roofing and Siding Co. Escanaba, MI	1984	2	1930
Jayhawk Roofing Co., Inc. Salina, KS	1977	4	1930
Joyce and Kramer Inc. Albany, NY	1972	2	1930

	Joined NRCA	No. of Generations	Founded
Montgomery Tin Shop Inc. Montgomery, AL	1985	–	1930
Curran V. Nielsen Co., Inc. Minneapolis, MN	1974	–	1930
Petrolane Long Beach, CA	1984	–	1930
Queen City Roofing and Contracting Co. Springfield, MO	1956	3	1930
Leo E. Richards Roofing and Sheet Metal Co. Louisville, KY	1968	3	1930
Harry E. Wendlandt Co., Inc. Appleton, WI	1975	2	1930
Bradford Roofing and Insulation Co. Billings, MT	1967	3	1931
California Roofing Co., Inc. San Jose, CA	1985	–	1931
Carpenter's Roofing and Sheet Metal Inc. West Palm Beach, FL	1969	–	1931
Fiberglas Canada Inc. Ontario, Canada	1984	–	1931
Goodburn Brothers Inc. Columbus, OH	1981	3	1931
Greenwood Sheet Metal Works Inc. Greenwood, MS	1980	3	1931
Highwood Roofing Co. Tenafly, NJ	1979	–	1931

	Joined NRCA	No. of Generations	Founded
A. T. Klemens and Son Great Falls, MT	1953	–	1931
Laco Contractors Inc. Falconer, NY	1985	–	1931
Lee County Metal and Roofing Co., Inc. Fort Myers, FL	1980	–	1931
Nordmann Roofing Co., Inc. Toledo, OH	1953	–	1931
Premier Roofing Co., Inc. West Haven, CT	1967	4	1931
Ramig Roofing Co., Inc. Wayne, NJ	1976	–	1931
Asbestos Roofing and Sheet Metal Co., Inc. Lima, OH	1966	2	1932
The Barrett Co. Millington, NJ	1981	4	1932
Clearview Equipment Co. St. Louis, MO	1985	–	1932
Empire Roofing and Insulation Co. Tulsa, OK	1955	–	1932
Ettel and Franz Co. St. Paul, MN	1953	–	1932
Garlock-French Roofing Corp. Minneapolis, MN	1968	1	1932
Gilsonite Corp. Portland, OR	1981	3	1932
Ora B. Hopper and Son Inc. Phoenix, AZ	1984	2	1932
Hou-Tex Roofing Co., Inc. Houston, TX	1972	–	1932

	Joined NRCA	No. of Generations	Founded
Iowa Falls Roofing Co. Iowa Falls, IA	1961	2	1932
Frank Kerby and Sons Inc. Waynesboro, VA	1980	2	1932
Lambert Roofing Co. Haverhill, MA	1984	3	1932
P. F. LaDuke and Son Roofing and Sheet Metal Detroit, MI	1960	–	1932
Liberty Roofing Co., Inc. Baltimore, MD	1967	2	1932
Al Melanson Co., Inc. Keene, NH	1963	–	1932
National Roofing and Siding Co. New Orleans, LA	1953	3	1932
New Brunswick Roofing and Metal Co. So. Plainfield, NJ	1981	–	1932
E. W. Olson Co., Inc. Markham, IL	1958	–	1932
Palmer Asphalt Co. Bayonne, NJ	1983	3	1932
Pawelko Frenzel Inc. Elk Grove, IL	1986	4	1932
Southern Roofing and Insulation Co., Inc. Augusta, GA	1973	–	1932
Robert Stiefel and Son Irvington, NJ	1985	–	1932
U.S. Industries Group Inc. Evansville, IN	1974	–	1932

	Joined NRCA	No. of Generations	Founded
Valley Roofing Co. Bay City, MI	1953	4	1932
Barrett-Nonpariel Roofing Inc. Norwalk, CT	1982	–	1933
Joseph S. Bowling Co., Inc. Louisville, KY	1958	3	1933
Consolidated Roofing and Supply Co. Phoenix, AZ	1973	–	1933
H. C. and E. F. Gilbert Inc. Williamsville, NY	1972	3	1933
Hamilton Roofing Co. Lubbock, TX	1967	2	1933
Haws Roofing Co., Inc. Tyler, TX	1980	3	1933
Henry Co. Huntington Park, CA	1970	2	1933
Michigan Roofing Co. Detroit, MI	1970	–	1933
Southwestern Petroleum Corp. Fort Worth, TX	1974	–	1933
Standard Roofings Inc. Tinton Falls, NJ	1981	–	1933
Bryant Organization Inc. Carson, CA	1953	3	1934
Greenstreak St. Louis, MO	1986	2	1934
J. P. Lahr and Sons Roofing Service Indianapolis, IN	1975	3	1934
Ledbetter Roofing Co. Atlanta, GA	1961	2	1934
Meijer Inc. Grand Rapids, MI	1983	3	1934
Preformed Roof Decks Inc. Kansas City, MO	1975	1	1934
Roofing Consultants and Inspection Edmonds, WA	1986	–	1934
Safeway Roofing and Siding Co. San Francisco, CA	1981	–	1934
Stone Roofing Co., Inc. Azusa, CA	1980	–	1934
West Georgia College Carrollton, GA	1983	–	1934
Whatley Roofing and Sheet Metal Inc. Tampa, FL	1979	2	1934
Arco Chemical Co. Philadelphia, PA	1975	–	1935
Barger-Ashe Roofing Co. Lenoir, NC	1953	3	1935
Brady Roofing and Sheet Metal Inc. Miami, FL	1972	2	1935
G. Brouillette and Son Inc. Raynham, MA	1972	–	1935
Cyclone Roofing Co. Matthews, NC	1981	2	1935
Edwards Roofing Co., Inc. Pensacola, FL	1975	3	1935
Industrial Cork Co., Inc. Elmhurst, IL	1973	3	1935

	Joined NRCA	No. of Generations	Founded
Jack's Roofing Co., Inc. Bethesda, MD	1976	2	1935
Johnson Roofing Inc. Waco, TX	1979	2	1935
Chas. Magid's Sons Inc. Long Island City, NY	1982	–	1935
W. B. Maske Sheet Metal Works Inc. Bladensburg, MD	1982	2	1935
Paulsen Roofing Inc. Salt Lake City, UT	1953	2	1935
Pine Roofing Co. Chicago, IL	1958	–	1935
Sechrist-Hall Co. Corpus Christi, TX	1953	–	1935
Simpson Metal Industries Inc. College Point, NY	1981	–	1935
Texas Roofing Co. Lubbock, TX	1953	–	1935
Trumbull Asphalt Summit, IL	1971	–	1935
M. Weisman Roofing Co., Inc. Cranston, RI	1985	3	1935
Babel Roofing Co. Detroit, MI	1986	3	1936
Clements Roofing Inc. Chicago, IL	1958	3	1936
Cumberland Roofers Inc. Valley Stream, NY	1980	2	1936
Elmer W. Davis Inc. Rochester, NY	1967	3	1936

	Joined NRCA	No. of Generations	Founded
Fort Roofing and Sheet Metal Works Sumter, SC	1953	2	1936
Glossop Roofing and Siding Inc. Aurora, IL	1974	3	1936
Harris Brothers Roofing Co. Charleston, WV	1956	–	1936
Industrial First Inc. Columbus, OH	1958	1	1936
Industrial First Inc. Cleveland, OH	1968	1	1936
R. C. Kimbrel Roofing Co. Elk Grove, IL	1965	3	1936
Ernest Peterson Inc. Hartford, CT	1972	4	1936
Potts Construction and Roofing Inc. Sedalia, MO	1984	2	1936
C. E. Reeve and Sons Inc. Indianapolis, IN	1953	4	1936
Rike Roofing and Manuf. Co., Inc. Charlotte, NC	1974	–	1936
J. L. Robbins Co. Cleveland, OH	1981	–	1936
Service Roofing Co. Waterloo, IA	1978	–	1936
Technicote Corp. Memphis, TN	1980	–	1936
Vincent Metals Minneapolis, MN	1974	–	1936

	Joined NRCA	No. of Generations	Founded
Weldon Roofing and Sheet Metal Inc. Weldon, NC	1953	2	1936
Wimsatt Brothers Inc. Louisville, KY	1983	3	1936
Alkor Division of Hedwin Corp. Pasadena, TX	1980	–	1937
Associated Roofing and Insulation Redwood City, CA	1984	2	1937
Central Oregon Roofing Inc. Bend, OR	1977	2	1937
M. Downes and Co. McKeesport, PA	1978	3	1937
George H. Duross Inc. Philadelphia, PA	1953	–	1937
Hartman Roofing and Sheet Metal Co. Cleveland, OH	1970	3	1937
Edward J. Laperouse Metal Works Inc. Houma, LA	1977	–	1937
Lunday Thagard Co. South Gate, CA	1984	–	1937
Hugh McNiven Co. Seattle, WA	1981	–	1937
Modern Materials Inc. Irvine, CA	1976	3	1937
Southern Wisconsin Roofing Co., Inc. Madison, WI	1972	2	1937
A. W. Therrien Co., Inc. Manchester, NH	1958	3	1937

	Joined NRCA	No. of Generations	Founded
Valdosta Roofing and Supply Co., Inc. Valdosta, GA	1986	1	1937
Vulcan Roofing and Sheet Metal Co., Inc. Birmingham, AL	1967	–	1937
Acme Roofing and Sheet Metal Co. Bloomington, IL	1980	–	1938
Cookeville Sheet Metal Works Inc. Cookeville, TN	1967	2	1938
Ferber Sheet Metal Works Inc. Jacksonville, FL	1966	2	1938
The Flat Roofing Contractors Advisory Board West Sussex, England	1982	–	1938
Johnson-Hilliard Inc. Kingsport, TN	1972	2	1938
Mayo Roofing Inc. Concord, NH	1982	1	1938
Mays Roofing Inc. Kokomo, IN	1975	–	1938
Smith-Graham Roofing Co. Battle Creek, MI	1956	2	1938
Statesville Roofing and Heating Co., Inc. Statesville, NC	1974	2	1938
Valley Sheet Metal Works Inc. Middletown, OH	1959	2	1938
Western Roofing Co., Inc. Bellingham, WA	1978	2	1938

	Joined NRCA	No. of Generations	Founded
W. J. Woodruff Roofing Contractors Fond du Lac, WI	1957	3	1938
W. J. Wray Contractors Inc. Gaston, NC	1985	3	1938
A. L. L. Roofing and Building Materials Corp. Los Angeles, CA	1981	3	1939
Burlington Roofing Co., Inc. Burlington, NC	1953	2	1939
Consolidated Fiberglass Products Co. Bakersfield, CA	1980	4	1939
Ganser Roofing Systems Inc. Madison, WI	1981	–	1939
Gates Engineering Co., Inc. Wilmington, DE	1976	–	1939
The Harrod Co., Inc. Kendallville, IN	1974	2	1939
Home Roofing and Building Co. Murfreesboro, TN	1969	3	1939
R. Kaller and Sons Ardmore, PA	1984	3	1939
Joseph Miorelli and Co., Inc. Hazleton, PA	1974	–	1939
Monsey Products Co. Kimberton, PA	1981	–	1939
Rock-Tred Corp. Skokie, IL	1973	–	1939
Major L. Rodd St. Johnsbury, VT	1968	3	1939

	Joined NRCA	No. of Generations	Founded
Seeback and Sons (1979) Ltd. Ontario, Canada	1982	–	1939
Joe Summers Roofing Co. Chamblee, GA	1973	–	1939
Vaughan Roofing Co., Inc. Baton Rouge, LA	1979	2	1939
Yancey Co. Sacramento, CA	1968	2	1939

1940–1965

	Joined NRCA	No. of Generations	Founded
Airport Roofing Co., Inc. Bridgeton, MO	1984	–	1940
Anning Johnson Co. Melrose Park, IL	1982	–	1940
Chaffee Roofing Co. Ferndale, MI	1985	4	1940
Gulf States Asphalt Co., Inc. Houston, TX	1981	–	1940
Leigh Coopersville, MI	1982	–	1940
D.H Mayou Roofing and Supply Co., Inc. Ottawa, IL	1953	3	1940
Richland Co. and Associates Inc. Defiance, OH	1986	3	1940
Squires Construction Co. Macedonia, OH	1983	–	1940
Adams and Sons Ltd. Janesville, WI	1974	2	1941
Alpek Sheet Metal and Roofing Rochester, MN	1982	3	1941

	Joined NRCA	No. of Generations	Founded
Colonial Roofing Co., Inc. West Haven, CT	1984	3	1941
Complete Roofing Co., Inc. Imperial, MO	1985	–	1941
Frontier Roofing Co. of El Paso El Paso, TX	1967	3	1941
Greenville Roofing Co. Greenville, SC	1970	–	1941
Hanson Roofing Inc. Evanston, IL	1986	2	1941
Hunt Roofing Co. Petoskey, MI	1983	–	1941
Law Engineering Testing Co. Atlanta, GA	1983	–	1941
Lough Brothers Roofing and Siding Co., Inc. Terre Haute, IN	1981	2	1941
Loyal Roofing Co. Chicago, IL	1985	2	1941
Rubber and Plastics Compound Co. Long Island City, NY	1973	2	1941
Fabco Fastening Systems Stanfield, NC	1979	–	1942
Peterson Roofing Co., Inc. Olean, NY	1974	3	1942
Pinkston-Hollar Inc. Arlington, TX	1980	3	1942
Universal Roofing and Sheet Metal Co., Inc. New Bedford, MA	1972	2	1942

	Joined NRCA	No. of Generations	Founded
Atlas Roofing and Supply Co. Dallas, TX	1969	2	1943
T. F. Beck Co. Auburn Heights, MI	1965	3	1943
Brattleboro Roofing and Sheet Metal Brattleboro, VT	1981	–	1943
Clingan Roofing-Siding-Insulation Co. Ludington, MI	1975	–	1943
Hays Roofing and Supply Inc. Phoenix, AZ	1984	3	1943
J. N. Vail Co., Inc. Wenatchee, WA	1974	3	1943
Valley Roofing Corp. Roanoke, VA	1973	–	1943
Vincent Roofing Inc. Topeka, KS	1953	2	1943
Voegele Co., Inc. Pittsburgh, PA	1953	2	1943
Allied Industries Houston, TX	1967	–	1944
Bernard L. Dalsin Co. Minneapolis, MN	1958	3	1944
Empire Roofing Co. Chicago, IL	1953	3	1944
J and J Roofing and Sheet Metal Co., Inc. Dallas, TX	1978	2	1944
Jones Brothers Roofing Co., Inc. Montgomery, AL	1966	3	1944

	Joined NRCA	No. of Generations	Founded
I. N. McNeil Roofing and Sheet Metal Inc. Roanoke, VA	1983	2	1944
Monahan and Loughlin Inc. Plattsburgh, NY	1974	3	1944
Thomas D. Robison Roofing Inc. Blackfoot, ID	1980	–	1944
Stan the Roof Man Inc. South San Francisco, CA	1986	2	1944
Tamko Asphalt Products Inc. Joplin, MO	1973	3	1944
L. W. Thomas Roofing Inc. Peoria, IL	1955	–	1944
Alumax Mesquite, TX	1981	–	1945
Andrews Roofing and Sheet Metal Inc. Carroll, IA	1957	1	1945
Blindauer's Inc. Green Bay, WI	1984	2	1945
Borsche Roofing Co., Inc. Appleton, WI	1977	2	1945
Carlson Associates Mableton, GA	1984	–	1945
The Carlson Group Cochituate, MA	1975	–	1945
Consolidated Protective Coatings Corp. Cleveland, OH	1981	–	1945
C. A. Eckstein Inc. Cincinnati, OH	1979	1	1945

	Joined NRCA	No. of Generations	Founded
The Ellis Co. Sacramento, CA	1959	–	1945
N. B. Emory Roofing Inc. Greensboro, NC	1971	3	1945
Chas. F. Evans Co., Inc. Elmira, NY	1970	2	1945
Gooding Simpson and Mackes Inc. Ephrata, PA	1969	2	1945
Greenberg Roofing Co. Grand Forks, ND	1984	–	1945
W. P. Hickman Co. Asheville, NC	1977	2	1945
E. L. Hilts and Co. Hickory, NC	1964	–	1945
Ibos Roofing Co., Inc. Covington, LA	1978	–	1945
Lawmaster Brothers Inc. Bourbon, IN	1958	1	1945
Marshall Roofing and Sheet Metal Co. Broken Arrow, OK	1984	4	1945
B. B. McCormick Roofing Co. Orlando, FL	1970	3	1945
Midwest Roofing-Sheet Metal Evansville, IN	1954	–	1945
Nieman Roofing Co., Inc. New Prague, MN	1981	2	1945
Pickard Roofing Co., Inc. Durham, NC	1981	3	1945
TOPCOAT Inc. Walpole, MA	1985	–	1945

	Joined NRCA	No. of Generations	Founded
United Roofing Los Angeles, CA	1982	–	1945
Acme Roofing Co. of Austin Austin, TX	1971	3	1946
Adams and Beagles Roofing Co., Inc. Hialeah, FL	1980	3	1946
Beldon Roofing and Remodeling Co. San Antonio, TX	1972	3	1946
William H. Byars Roofing Co. Ontario, CA	1982	2	1946
Clark Roofing Co. Broadview, IL	1974	2	1946
Colonial Roofing Co., Inc. College Point, NY	1981	3	1946
Colonial Rubber Works Inc. Dyersburg, TN	1982	–	1946
Consumers/Dornin-Adams Inc. Lynchburg, VA	1962	–	1946
Crafts Inc. Manitowoc, WI	1959	3	1946
Crow Roofing and Sheet Metal Inc. Seattle, WA	1965	2	1946
Embrey Roofing Co. San Antonio, TX	1979	3	1946
A. W. Farrell and Son Dunkirk, NY	1967	–	1946
Federal Sheet Metal and Roofing Co. Jamaica Plain, MA	1960	–	1946
Field and Associates Inc. Springfield, OH	1957	2	1946
Frost and Co., Inc. Wapakoneta, OH	1979	3	1946
Gastonia Sheet Metal Works Inc. Gastonia, NC	1980	–	1946
Gilbert and Becker Co., Inc. Boston, MA	1966	–	1946
Hub Sheet Metal and Roofing Inc. Hattiesburg, MS	1978	3	1946
J and P Petroleum Products Dallas, TX	1983	2	1946
L. P. Kent Corp. Bronx, NY	1979	2	1946
Likar Roofing Co., Inc. North Versailles, PA	1980	–	1946
Lee M. Machemer and Son Allentown, PA	1985	3	1946
MacArthur Roofing and Sheet Metal Co. Saginaw, MI	1961	3	1946
Mandal's Roofing and Waterproofing Contractors Gulfport, MS	1977	–	1946
McGonigle and Hilger Roofing Inc. Lockport, NY	1975	–	1946
Nick Michels and Sons Inc. Nekoosa, WI	1955	3	1946
Middleton Roofing Co. Mentor, OH	1982	2	1946

	Joined NRCA	No. of Generations	Founded
Midwest Sales Co. St. Louis, MO	1983	–	1946
Donald M. Miller Roofing Co. Uniontown, PA	1980	2	1946
Montgomery Roofing Co., Inc. Miami, FL	1979	1	1946
Nixon Roofing Inc. Fredericksburg, TX	1978	–	1946
Reimann and Georger Inc. Buffalo, NY	1966	–	1946
Rollins Supply Co., Inc. Greensboro, NC	1974	2	1946
San Diego Roofing Co., Inc. National City, CA	1979	–	1946
Universal Sheet Metal and Roofing Co. Providence, RI	1976	2	1946
Wheeler Roofing Service Inc. Columbus, MS	1978	2	1946
Wilkinson Roofing and Siding Inc. Wilmington, DE	1982	–	1946
Abernathy and Clark Sheet Metal and Roofing Inc. Tupelo, MS	1980	2	1947
Ajax Roofing Co. Lubbock, TX	1979	2	1947
Allentown Roofing and Sheet Metal Co., Inc. Allentown, PA	1972	2	1947
American Building Co. Eufaula, AL	1983	–	1947

	Joined NRCA	No. of Generations	Founded
B and M Roofing of Boulder Inc. Boulder, CO	1979	2	1947
Babb Sheet Metal Co. Wilmington, OH	1981	–	1947
Burris Building Materials Dallas, TX	1986	–	1947
Daly Protective Coatings Co., Inc. Hammond, IN	1982	3	1947
Elgin Roofing Co. Elgin, IL	1983	–	1947
EG and G Florida Inc. Orlando, FL	1986	–	1947
Ferguson Roofing and Supply Inc. Lake Charles, LA	1980	2	1947
General Roofing and Siding Port Huron, MI	1982	2	1947
Geoghegan Corp. Louisville, KY	1977	3	1947
E. C. Goldman Inc. Winter Park, FL	1960	–	1947
W. J. Grinder Roofing Co., Inc. Rochester, NY	1980	1	1947
Gulf Waterproofing Co., Inc. Houston, TX	1967	–	1947
Haug Roofing Inc. West Bend, WI	1977	–	1947
Clovis Hendry Industries Inc. Addis, LA	1974	–	1947
R. D. Herbert and Sons Co. Nashville, TN	1965	3	1947

	Joined NRCA	No. of Generations	Founded
Hinshaw Roofing and Sheet Metal Co., Inc. Frankfort, IN	1957	2	1947
Holt Roofing Co., Inc. Toledo, OH	1985	3	1947
Honolulu Roofing Co., Ltd. Honolulu, HI	1960	–	1947
Hurley Construction and Roofing Co., Inc. Somerville, MA	1981	3	1947
Industrial Roofing and Asbestos Co. Youngstown, OH	1973	–	1947
Industrial Roofing and Sheet Metal Works Inc. Shreveport, LA	1980	–	1947
Lane Roofing Co. Lake City, SC	1984	–	1947
L. R. Lloyd Co. Uniontown, PA	1969	2	1947
Ben J. Malone Co. Memphis, TN	1958	–	1947
Marlette Roofing and Sheet Metal Co. Marlette, MI	1974	2	1947
Clay McCullough and Co., Inc. San Jose, CA	1985	–	1947
Modern Roof and Insulation Co., Inc. Pocatello, ID	1973	–	1947
Parsley's Sheet Metal and Roofing Co., Inc. Pampa, TX	1978	3	1947
Peninsula Roofing Co. Salisbury, MD	1970	2	1947
Dave Pomaville and Sons Inc. Warren, MI	1967	3	1947
Republic Powdered Metals Inc. Medina, OH	1979	–	1947
Samaritano and Co., Inc. San Juan, PR	1958	–	1947
Skyway Roofing of Troy Inc. Troy, NY	1978	2	1947
Smith Roofing Co., Inc. Brownwood, TX	1984	3	1947
Stevenson Co., Inc. Topeka, KS	1977	–	1947
S. E. Sulenski Roofing and Siding Holyoke, MA	1969	–	1947
Trion Inc. Sanford, NC	1984	–	1947
United Roofing and Construction Inc. Laurel, MS	1977	3	1947
Beacon Roofing Co., Inc. Beacon, NY	1974	–	1948
Boone Brothers Roofing Inc. Omaha, NE	1984	3	1948
Bowman Roofing and Sheet Metal Co. Hickory, NC	1975	3	1948
Caldwell-Roland Roofing Co., Inc. Berkeley, CA	1980	2	1948
Dolph Roofing Co., Inc. Dallas, TX	1982	1	1948

	Joined NRCA	No. of Generations	Founded
Earl F. Douglass Roofing Co. Greely, CO	1953	–	1948
Fidelity Roof Co. Oakland, CA	1970	2	1948
R. E. Forshee Co., Inc. Cincinnati, OH	1963	2	1948
Fowler and Peth Inc. Denver, CO	1973	2	1948
Gory Associated Industries Inc. North Miami, FL	1972	–	1948
Henschel Roofing Co. Bismarck, ND	1985	2	1948
Holliday Roofing Co., Inc. Tyler, TX	1979	3	1948
Johnson-Vose Inc. Tucson, AZ	1960	2	1948
Naturalite Inc. Garland, TX	1976	2	1948
Ontario Building Supply Co., Inc. Rochester, NY	1986	–	1948
Pyramid Roofing Co., Inc. Sikeston, MO	1973	2	1948
R and R Industries Inc. Daytona Beach, FL	1972	2	1948
Donald B. Smith Inc. Hanover, PA	1981	–	1948
Stanislaus Roofing Inc. Modesto, CA	1979	2	1948
Watersaver Co., Inc. Denver, CO	1981	2	1948

	Joined NRCA	No. of Generations	Founded
Zerbe Roofing Co. Modesto, CA	1984	–	1948
Cradock Sheet Metal and Roofing Co., Inc. Portsmouth, VA	1983	2	1949
Dexter Mook Roofing and Sheet Metal Work's Inc. Chicago, IL	1975	–	1949
The Fick Brothers Roofing Co. Baltimore, MD	1972	2	1949
Firebaugh and Reynolds Roofing Co. Novi, MI	1964	3	1949
Guilford Mills Inc. Greensboro, NC	1984	–	1949
Earl Hankins Roofing Co. Kansas City, MO	1974	–	1949
Korb Roofers Inc. Baltimore, MD	1968	2	1949
G. A. Largent and Associates Inc. Cumberland, MD	1976	–	1949
Le-Ray Roofing Co., Inc. Streator, IL	1963	2	1949
Morgan Roofing Co., Inc. Lake Charles, LA	1976	2	1949
Perlite Institute Inc. Commack, NY	1975	–	1949
D. C. Taylor Co. Cedar Rapids, IA	1977	3	1949
Weaver Metal and Roofing Co., Inc. Buffalo, NY	1956	2	1949

	Joined NRCA	No. of Generations	Founded
Western Pacific Roofing Corp. Lancaster, CA	1969	–	1949
Aquaproof Roofing Co., Inc. Bradenton, FL	1979	2	1950
Harold J.Becker Co., Inc. Dayton, OH	1972	–	1950
Berwald Roofing Co. North St. Paul, MN	1981	2	1950
Carl's Roofing and Siding Co., Inc. Tampa, FL	1974	2	1950
S. D. Carruthers Sons Inc. Argyle, NY	1982	3	1950
Construction Fasteners Inc. Wyomissing, PA	1979	2	1950
Delta Roofing and Sheet Metal Corp. West Memphis, AR	1978	–	1950
Dun-Rite Shingle Co., Inc. Richmond Hill, NY	1986	–	1950
East Muskegon Roofing and Sheet Metal Muskegon, MI	1967	2	1950
J. S. Filo Jr. Inc. Plainfield, NJ	1978	1	1950
Dale R. Horning Co., Inc. Indianapolis, IN	1955	–	1950
Krueger Sheet Metal Co. Spokane, WA	1980	–	1950
Marty Kuzmkowski Roofing Delmont, PA	1982	2	1950
E. H. Lang Roofing Inc. Cortland, NY	1978	1	1950

	Joined NRCA	No. of Generations	Founded
Don Largent Roofing Inc. Harrisonburg, VA	1969	3	1950
Milbrand Co. Warren, MI	1967	2	1950
Pettigrew and Sons Fort Wayne, IN	1967	2	1950
Morgen Manufacturing Co. Yankton, SD	1977	1	1950
Romig Roofing Co. St. Clairsville, OH	1953	2	1950
Ross Roofing Co. Sand City, CA	1981	2	1950
Rupo Technical Services Oak Creek, WI	1985	2	1950
Southern Maryland Alum. Prod. Co. Davidsonville, MD	1980	2	1950
MacDonald Sprague Roofing Co., Inc. Holbrook, MA	1967	–	1950
Tampa Roofing Co. Tampa, FL	1970	3	1950
Tuckahoe Metal and Roofing Inc. Yonkers, NY	1975	–	1950
Youngs Sheet Metal and Roofing Inc. Asheville, NC	1979	2	1950
Alice Roofing and Sheet Metal Works Inc. San Antonio, TX	1976	–	1951
Anderson Sheet Metal Works Inc. Winchester, VA	1976	2	1951

	Joined NRCA	No. of Generations	Founded
B and F Supply Co., Inc. Daytona Beach, FL	1979	–	1951
Bonded Applicators of Maryland Inc. Hagerstown, MD	1974	2	1951
Joe Clesson's Roofing Inc. Tacoma, WA	1968	1	1951
Dale Crampton Co. Fort Smith, AR	1955	–	1951
Dollar Roofing Co. Goleta, CA	1974	–	1951
DTK Co., Inc. Redlands, CA	1979	–	1951
Gentry Industrial Service Dublin, IN	1983	1	1951
Guardian Roofing Systems West Caldwell, NJ	1986	2	1951
Holland Ready Roofing Co. Holland, MI	1974	–	1951
Hurlock Roofing Co. Wilmington, DE	1973	–	1951
Knox Gill Co. Little Rock, AR	1971	–	1951
L and L Roofing and Construction El Paso, TX	1984	2	1951
James A. McBrady Inc. Portland, ME	1964	2	1951
Modern Kraft Roofing Co., Inc. Whittier, CA	1976	2	1951
Newman Wholesale Inc. Asheville, NC	1983	–	1951

	Joined NRCA	No. of Generations	Founded
Peterson Brothers Roofing St. Paul, MN	1985	1	1951
Roofers Inc. Baltimore, MD	1970	2	1951
Royal Roofing Co. Oak Park, MI	1967	–	1951
Stephenson Roofing and Sheet Metal Co. Maryland Heights, MO	1958	2	1951
Sunflower Roofing and Industries Inc. El Dorado, KS	1971	3	1951
United Roofing Co., Inc. Fort Dodge, IA	1974	–	1951
Waukegan Roofing Co., Inc. Waukegan, IL	1956	2	1951
Winpigler Roofing Inc. Frederick, MD	1984	2	1951
Wright-Brown Roofing Co. Detroit, MI	1967	2	1951
Airlite Processing Corp. Vero Beach, FL	1982	–	1952
Ballard Roofing Co., Inc. Livonia, MI	1967	2	1952
J. W. Conway Inc. Hyattsville, MD	1974	–	1952
James Cox and Son Inc. Baltimore, MD	1969	–	1952
Dalbec Roofing and Sheet Metal Inc. Long Lake, MN	1984	–	1952

	Joined NRCA	No. of Generations	Founded
Dunmore Roofing and Supply Co., Inc. Dunmore, PA	1978	2	1952
Embassy Roofing Inc. Chicago, IL	1976	–	1952
Eversole Co. Dallas, TX	1986	2	1952
General Roofing and Tile Co., Inc. St. Petersburg, FL	1966	3	1952
Hathcock Roofing and Remodeling Co., Inc. Dothan, AL	1981	–	1952
Hedinger's Roofing Co., Inc. Jasper, IN	1979	–	1952
Hershey Roofing and Sheet Metal Inc. Albert Lea, MN	1975	–	1952
Linwood Roofing and Contracting Brooklyn, NY	1981	3	1952
Sapp Roofing Co., Inc. Pittsburgh, PA	1967	1	1952
Scholten Roofing Inc. Lynden, WA	1982	2	1952
Seppala and Aho Roofing Division New Ipswich, NH	1980	1	1952
Southern Roofing Co., Inc. Jackson, MS	1979	3	1952
Weather Control Co. Montoursville, PA	1984	–	1952
Whitehead Roofing and Insulation Inc. Belleville, IL	1973	–	1952

	Joined NRCA	No. of Generations	Founded
Barrett Roofing and Supply Co. of Danbury Danbury, CT	1980	–	1953
Ernest Braman Roofing Hillsdale, MI	1966	–	1953
Brown Roofing and Sheet Metal Greenville, MS	1977	–	1953
El Pueblo Roofing Co. Tucson, AZ	1984	3	1953
Elastizell Corp. of America Ann Arbor, MI	1975	2	1953
Enichem Americas New York, NY	1986	–	1953
ES Products Inc. New Rochelle, NY	1971	1	1953
Fresno Roofing Co. Fresno, CA	1968	–	1953
Harwood Roofing Inc. Salt Lake City, UT	1981	3	1953
Industrial Metal Fabricators Inc. Cleveland, OH	1967	–	1953
John's Roofing and Sheet Metal Co., Inc. Toccoa, GA	1980	2	1953
Kiker Roofing Corp. Pleasantville, NJ	1978	1	1953
W. A. Lynch Roofing Co., Inc. Charlottesville, VA	1973	2	1953
Middle Tenn. Roofing Co., Inc. Nashville, TN	1984	1	1953
J. C. Mitchell and Son Roseberg, OR	1984	1	1953

	Joined NRCA	No. of Generations	Founded
Modern Method Roofing Co., Inc. Napa, CA	1981	–	1953
Orndorff and Spaid Inc. Beltsville, MD	1971	2	1953
Oyster Bay Roofing and Sheet Metal Inc. Oyster Bay, NY	1971	–	1953
Patten Roofing Co. Winnebago, MN	1968	–	1953
J. F. Probst and Co., Inc. Butler, WI	1955	–	1953
Reserve Roofing and Sheet Metal Inc. Akron, OH	1968	1	1953
Roofmaster Products Co. Los Angeles, CA	1963	2	1953
Tilsen Roofing Co. Madison, WI	1959	–	1953
Tolley-Hughes Inc. Boise, ID	1976	–	1953
Albert Roofing Co., Inc. High Point, NC	1961	–	1954
Anderson Roofing Co., Inc. Portland, OR	1972	–	1954
Beyer Roofing Co., Inc. Saginaw, MI	1981	–	1954
Bonitz Insulation Co. Columbia, SC	1982	–	1954
Brown Roofing Co. The Dalles, OR	1975	–	1954
A. F. Callan and Co., Inc. Palmyra, NJ	1986	–	1954
Cantrell Sheet Metal and Roofing Co., Inc. Milan, TN	1981	–	1954
Deeringer and Son Inc. Lexington, KY	1978	2	1954
Fulton Roofing Co. Atlanta, GA	1979	1	1954
Mobay Chemical Corp. Pittsburgh, PA	1976	–	1954
Model Construction and Supply Knoxville, TN	1985	–	1954
Ohio Building Restoration Inc. Toledo, OH	1986	–	1954
J. W. Reneau Roofing Co. Houston, TX	1982	2	1954
Southwestern Roofing and Metal Co. Oklahoma City, OK	1955	1	1954
Superior Roofing Co., Inc. Salt Lake City, UT	1973	2	1954
Wayne's Roofing and Sheet Metal Co., Inc. Daytona Beach, FL	1984	–	1954
Allstate Roofing Inc. Billings, MT	1979	3	1955
Bainbridge Sheet Metal Works Inc. Bainbridge, GA	1980	3	1955
Bright Roofing and Sheet Metal Inc. Detroit, MI	1973	–	1955

	Joined NRCA	No. of Generations	Founded
Campbell's Roofing and Sheet Metal Inc. Dayton, OH	1978	–	1955
Causey Roofing Corp. Miami, FL	1979	3	1955
Commercial Roofing and Sheet Metal Co. Columbia, SC	1969	3	1955
Crow Roofing Inc. Syracuse, IN	1979	–	1955
L. J. Crowther Co. Joliet, IL	1962	3	1955
Cudahy Roofing and Supply Inc. Cudahy, WI	1970	–	1955
Dallas Wholesale Builders Supply Inc. Dallas, TX	1983	–	1955
Early Roofing Service Inc. Springfield, OR	1981	1	1955
Holmes Associates Inc. Farmington Hills, MI	1979	2	1955
Indiana Supply Corp., Inc. Indianapolis, IN	1981	2	1955
Kitson Brothers Inc. Hatfield, PA	1978	–	1955
National Roofing Inc. Millington, NJ	1972	2	1955
Quality Roofers Inc. Twin Falls, ID	1985	3	1955
Standard Roofing and Sheet Metal Works Inc. Lake Worth, FL	1979	2	1955

	Joined NRCA	No. of Generations	Founded
Thermo Materials Inc. San Diego, CA	1973	–	1955
Wilson Enterprises Inc. Bensenville, IL	1974	2	1955
A and B Equipment Co. Fort Worth, TX	1978	–	1956
Asbestos Engineering and Supply Co., Inc. Phoenix, AZ	1980	–	1956
B and S Roofing Co., Inc. Pawtucket, RI	1984	1	1956
Bade Roofing and Sheet Metal Co. St. Louis, MO	1968	2	1956
Burns and Scalo Roofing Co. Bridgeville, PA	1983	–	1956
Easley Roofing and Sheet Metal Inc. Victoria, TX	1966	2	1956
Grove Roofing and Siding Co. Ivyland, PA	1977	–	1956
Hamilton and Spiegel Inc. Bladensburg, MD	1981	–	1956
Henris Supply Inc. Petaluma, CA	1968	1	1956
Holbrook Roofing Co., Inc. Gastonia, NC	1983	2	1956
William H. Kelly Co. Novi, MI	1969	2	1956
L and P Building Supply Albuquerque, NM	1982	2	1956
Simon Lono Ltd. Newfoundland, Canada	1971	1	1956

	Joined NRCA	No. of Generations	Founded
William C. McCombs and Co., Inc. Rochester, NY	1983	2	1956
Metal Distributors Inc. Columbia, SC	1984	1	1956
National Roofing Corp. Brooklyn, NY	1985	–	1956
Quad City Roofing Co. Madison, IL	1973	2	1956
Rash Roofing Inc. Houston, TX	1983	1	1956
Leonard Smith Sheet Metal and Roofing Inc. Salem, VA	1969	2	1956
South Roofing Corp. South Boston, MA	1977	–	1956
Ray St. Clair and Son Roofing Cincinnati, OH	1983	–	1956
Tozai Asphalt Roofing Contractors Co-op. Tokyo, Japan	1985	–	1956
Wiss-Janney-Elstner Associates Bloomington, IL	1983	–	1956
Advance Roofing Co. Phoenix, AZ	1973	2	1957
Beck Roofing Co., Inc. Hayward, CA	1980	2	1957
Coast Roof Co., Inc. Fullerton, CA	1981	–	1957
Donelson Roofing Co., Inc. Nashville, TN	1981	3	1957
Edmonds Roofing Co., Inc. Edmonds, WA	1984	1	1957

	Joined NRCA	No. of Generations	Founded
Frazier Roofing and Sheet Metal Co., Inc. Dyersburg, TN	1979	–	1957
Glenrock Co. Northlake, IL	1986	–	1957
Hall Roofing and Sheet Metal Co., Inc. Elmira, NY	1974	2	1957
The Hamlin Cos. Garner, NC	1967	–	1957
Krupnik Brothers Inc. Glen Burnie, MD	1967	–	1957
Livingston Roofing and Sheet Metal Co., Inc. Baton Rouge, LA	1967	–	1957
Ralph J. Meyer Co. Pittsburgh, PA	1982	–	1957
Miller-Thomas-Gyekis Inc. Pittsburgh, PA	1967	–	1957
Minter Roofing Co., Inc. Benton, KY	1979	1	1957
Ray Nolan Roofing Co., Inc. Louisville, KY	1961	–	1957
Pellerin and Wallace Inc. Lafayette, LA	1984	1	1957
Penn-Perry Inc. Wexford, PA	1983	–	1957
Rinehart Roofing Co. Topeka, KS	1981	2	1957
E. L. Scott Roofing Co., Inc. Kinston, NC	1974	–	1957

	Joined NRCA	No. of Generations	Founded
Security Roofing and Siding Co., Inc. Menasha, WI	1974	2	1957
Spann Inc. Myrtle Beach, SC	1968	2	1957
Stanley Roofing Miami, FL	1972	2	1957
Stark Roofing Co. Glenwood Springs, CO	1986	–	1957
Twin City Roofing and Sheet Metal Scottsbluff, NE	1980	3	1957
Vermont Roofing Rutland, VT	1964	–	1957
Ace Roofing Co. San Mateo, CA	1967	2	1958
Max Bayroff Corp. Bayonne, NJ	1976	2	1958
Cannon Roofing Co., Inc. Spartanburg, SC	1971	–	1958
Cascade Roofing Co. Portland, OR	1982	1	1958
Columbus College Columbus, GA	1984	–	1958
Dougherty Roofing Co. Lagrange, GA	1974	2	1958
Esmac Inc. Enterprise, AL	1978	–	1958
Howard Roofing Co. Las Vegas, NV	1977	–	1958
Local Roofing Co., Inc. Eau Claire, WI	1973	2	1958

	Joined NRCA	No. of Generations	Founded
L. R. Parsons Roofing Co., Inc. Great Bend, KS	1984	–	1958
Ray Roofing Co., Inc. West Palm Beach, FL	1969	2	1958
Roofing Wholesale Co., Inc. Phoenix, AZ	1973	2	1958
Richard L. Sensenig Co. Ephrata, PA	1978	–	1958
Singles Roofing Co., Inc. Elgin, IL	1981	2	1958
Skycraft Roofing Inc. Gardena, CA	1986	–	1958
Swanson Gentleman Inc. Des Moines, IA	1974	–	1958
Taylor Roofing Co., Inc. Topeka, KS	1959	2	1958
Thompson Roofing Co. Fargo, ND	1968	2	1958
Top's Roofing Co., Ltd. Wailuku, HI	1984	–	1958
Valley Wholesale West Springfield, MA	1963	–	1958
Bell's Roofing Co., Inc. Woonsocket, RI	1985	–	1959
Cole and Stevens Roofing Co. Fort Lauderdale, FL	1970	2	1959
Elens and Maichin Roofing and Sheet Metal Inc. Joliet, IL	1976	–	1959
Ertle Roofing and Sheet Metal Magnolia, NJ	1971	2	1959

	Joined NRCA	No. of Generations	Founded
F and F Roofing Co., Inc. Floral Park, NY	1979	–	1959
Galewood Tuckpointing and Roofing Co., Inc. Chicago, IL	1981	1	1959
Garlock Equipment Co. Minneapolis, MN	1968	3	1959
Owen E. Gupton Roofing Inc. Henderson, NC	1984	–	1959
Hatboro Roofing Co. Willow Grove, PA	1983	2	1959
Kelley and Carpenter Roofing and Sheet Metal Inc. Hamilton, OH	1978	–	1959
M J Roofing and Supply Ltd. Winnipeg, Canada	1984	–	1959
L. M. Martin Inc. Ephrata, PA	1985	–	1959
J. H. Mauldin Corp. San Antonio, TX	1982	2	1959
McLees Inc. Belgrade, MT	1980	–	1959
J. B. Passmore Roofing and Sheet Metal Co., Inc. Houston, TX	1972	–	1959
Roof Systems Inc. West Chester, PA	1973	3	1959
San Marino Roof Co., Inc. Orange, CA	1981	–	1959
Virginia Roofing Corp. Alexandria, VA	1963	–	1959

	Joined NRCA	No. of Generations	Founded
Wright Roofing Tacoma, WA	1983	2	1959
Adams County Roofing Inc. Commerce City, CO	1982	1	1960
Alpine Roofing Co., Inc. Denver, CO	1975	–	1960
American Roofing and Sheet Metal Co. Tulsa, OK	1980	–	1960
Brown Roofing and Sheet Metal Lamesa, TX	1973	2	1960
Buckaroo-Thermoseal Inc. Portland, OR	1969	–	1960
C and C Roofing Co. Chicago, IL	1984	–	1960
Construction Services Humble, TX	1982	–	1960
Construction Services Inc. Amarillo, TX	1973	3	1960
Eady Brothers Co., Ltd. Ontario, Canada	1985	–	1960
First Roofing and Sheet Metal Co. Lima, OH	1961	–	1960
Greer Roofing Inc. Greer, SC	1981	–	1960
Hausman Metal Works and Roofing Inc. St. Joseph, MO	1980	3	1960
Hi-Top Roofing and Coatings Inc. Orange, TX	1983	–	1960
Industrial Roofing Co., Inc. Winston-Salem, NC	1972	2	1960

	Joined NRCA	No. of Generations	Founded
J and L Roofing Co. Grand Rapids, MI	1969	–	1960
Kelly and Son Roofing Inc. Kansas City, KS	1973	2	1960
Kieff's Roofing Inc. Stanton, MI	1977	2	1960
Lexsuco Canada Ltd. Canada	1981	–	1960
Messersmith Roofing Inc. Cullman, AL	1980	1	1960
Mobay Corp. Pittsburgh, PA	1986	–	1960
Mr. Roof/Greaves Inc. Ypsilanti, MI	1986	–	1960
MM Systems Corp. Tucker, GA	1970	–	1960
Northwest Mono-Roof Inc. Spokane, WA	1985	2	1960
J. T. Penyak Roofing Co., Inc. South Plainfield, NJ	1984	–	1960
Regan Roofing Co., Inc. Mount Airy, NC	1967	–	1960
J. D. Rivet Co., Inc. Indian Orchard, MA	1978	2	1960
Robertson Roofing Co., Inc. Delray Beach, FL	1986	–	1960
R. L. Sanders Roofing Co. Smyrna, GA	1961	2	1960
Sidney Roofing Co., Inc. Sidney, NE	1981	2	1960
Sam S. Smucker and Sons Inc. Lancaster, PA	1980	2	1960
Thomas Roofing Co., Inc. Mobile, AL	1981	3	1960
Tri-State Roofing and Siding Hagerstown, MD	1974	–	1960
Ward Construction Inc. Danville, CA	1981	2	1960
Continental Roofing Contractors Inc. Utica, NY	1981	2	1961
D and S Roofing and Sheetmetal Corp. Worcester, MA	1966	–	1961
Design Containers Jacksonville, FL	1983	1	1961
Enterprise Roofing Service Concord, CA	1976	–	1961
Harvey Industries Inc. Waltham, MA	1984	–	1961
Russ Hines and Associates Inc. Auburn, NY	1974	2	1961
Jackson Roofing Co., Inc. Newark, NJ	1979	2	1961
L and L Roofing Co. Reno, NV	1969	1	1961
Martin Roofing Co., Inc. Wichita, KS	1973	–	1961
Mineral Fiber Manufacturing Corp. Coshocton, OH	1982	2	1961

	Joined NRCA	No. of Generations	Founded
Missouri Builders Service of Jefferson Cty Jefferson City, MO	1977	1	1961
National Roofing Laboratories Inc. Nashville, TN	1982	–	1961
Northwest Applicating Inc. Irving, TX	1985	–	1961
Prospect Industries Inc. Sterling, VA	1974	3	1961
Puget Sound Roofing Co. Seattle, WA	1973	1	1961
Sears Roofing Service Inc. Lakewood, CA	1973	1	1961
Summit BSR Roofing Bristol, TN	1979	3	1961
Tip Top Roofing and Sheet Metal Inc. Huntsville, AL	1970	2	1961
Chem-Wood Corp. St. Paul, MN	1984	–	1962
Complete Building Maintenance Co. Lombard, IL	1976	–	1962
Cribbs Inc. Baton Rouge, LA	1980	3	1962
Daniels Roofing Co. El Paso, TX	1973	3	1962
Foam-Crete Inc. Chattanooga, TN	1975	3	1962
James Green Roofing and Sheet Metal Co., Inc. Philadelphia, PA	1977	2	1962
Joseph Medas and Son Inc. Lexington, MA	1984	–	1962
Moppo Products Inc. Charlotte, NC	1975	2	1962
Mueller Roofing Service Inc. Phillipsburg, NJ	1979	–	1962
New Bedford Roofing and Sheet Metal Inc. New Bedford, MA	1976	1	1962
Nord Bitumi U.S. Inc. Springfield, NJ	1983	–	1962
Pioneer Roofing Co. Phoenix, AZ	1968	–	1962
R.S. Roofing and Sheet Metal Co., Inc. Nanuet, NY	1969	3	1962
Rain Proof Roofing Co., Inc. Anchorage, AK	1983	–	1962
Reeves Roofing Co., Inc. Van Nuys, CA	1985	2	1962
Silvercool Service Co. Denver, CO	1985	2	1962
Strickland Brothers Roofing Co., Inc. Dallas, TX	1980	3	1962
Sunset Roofing Co. San Francisco, CA	1985	–	1962
Wallace Roofing Co., Inc. Mobile, AL	1981	2	1962
Bentley Sheet Metal and Roofing Co., Inc. San Antonio, TX	1980	–	1963

	Joined NRCA	No. of Generations	Founded
Bridgeport Restoration Co., Inc. Bridgeport, CT	1979	–	1963
Deck Applicators Inc. Irving, TX	1975	–	1963
Durham Roofing Co., Inc. Durham, NC	1984	–	1963
Engineered Roofing Co. Sterling, IL	1963	3	1963
Paul J. Eyerman Inc. Nanticoke, PA	1968	2	1963
Fettin Roofing Co., Inc. Lincoln, NE	1980	2	1963
Giuffre Brothers Cranes Inc. Oak Creek, WI	1985	2	1963
Great Lakes Systems Inc. Jenison, MI	1968	2	1963
Henley Roofing Co. Knoxville, TN	1971	–	1963
C. R. Huffer Roofing Inc. Columbus, OH	1978	–	1963
Mansonville Plastic (BC) Ltd. Surrey, British Columbia	1985	–	1963
Maxwell Roofing and Sheet Metal Inc. Nashville, TN	1981	2	1963
McMurray and Sons Inc. Eureka, CA	1980	3	1963
Petrolane Gas Service Willow Springs, IL	1982	–	1963
Port Enterprises Inc. Port Lavaca, TX	1981	–	1963

	Joined NRCA	No. of Generations	Founded
Roofing Equipment and Spec. Co., Inc. Memphis, TN	1976	–	1963
Roofing Solutions Dayton, OH	1974	–	1963
Charles F. Ruff and Co., Inc. Baltimore, MD	1967	1	1963
Stephenson and Sons Roofing Inc. Flint, MI	1966	2	1963
T and F Systems Inc. Cleveland, OH	1974	–	1963
Barr Roofing Co. Abilene, TX	1973	1	1964
Commercial Roofing and Sheet Metal Co., Inc. Cheverly, MD	1979	3	1964
Crestline Supply Corp. Salt Lake City, UT	1982	2	1964
D J Inc. Brattleboro, VT	1979	–	1964
J. P. Fyfe Inc. Washington, NJ	1967	–	1964
Great Falls Roofing Inc. Great Falls, MT	1984	–	1964
HiLine Roofing Havre, MT	1981	–	1964
Kawkawlin Roofing Co. Kawkawlin, MI	1974	2	1964
Lincoln Property Co. Foster City, CA	1984	–	1964

	Joined NRCA	No. of Generations	Founded
Lucas Sales Co., Inc. St. Louis, MO	1974	1	1964
Martin Tomlinson Roofing Co. Dallas, TX	1984	–	1964
McKinney Roofing and Sheet Metal Inc. Austin, TX	1983	–	1964
National Nail Corp. Grand Rapid, MI	1984	–	1964
G. E. Riddiford Co. Arlington Heights, IL	1981	3	1964
RO Corp. Olathe, KS	1986	–	1964
U S G Industries Chicago, IL	1985	–	1964
Wendell Roofing Co. Cicero, IL	1976	1	1964
Ace Roofing Co., Inc. Glenwood Springs, CO	1982	–	1965
Ballard Roofing Inc. Maiden, NC	1983	2	1965
R. E. Burke Roofing Co., Inc. Skokie, IL	1970	2	1965
Carruth Roofing Co., Inc. Miami, FL	1969	–	1965
Clayton County BOE Maintenance Dept. Jonesboro, GA	1982	–	1965
Craft Roofing Corp. West Hempstead, NY	1981	–	1965
James D. Cummins and Co., Inc. Red Bank, NJ	1986	2	1965

	Joined NRCA	No. of Generations	Founded
Falls City Roofing Co., Inc. Jeffersonville, IN	1966	2	1965
Flame Engineering Inc. LaCrosse, KS	1985	2	1965
H. L. Gainey Roofing Co., Inc. Sumter, SC	1981	3	1965
General Builders Supply Corp. Fort Lauderdale, FL	1974	3	1965
Geoghegan Roofing and Supply Inc. Bowling Green, KY	1981	3	1965
The Haskell Co. Jacksonville, FL	1984	–	1965
Jones Roofing Co., Inc. San Jose, CA	1978	2	1965
Lloyd Roofing Co., Inc. Greenwood, SC	1980	2	1965
Mangrum Roofing Co., Inc. Mayfield, KY	1980	2	1965
Mississippi Valley Roofing Inc. Florissant, MO	1978	–	1965
Monier Roof Tile Orange, CA	1974	–	1965
Murr and Laney Inc. Charlotte, NC	1973	–	1965
O.K. Roofing Co., Inc. Laurens, IA	1972	2	1965
Petersen Aluminum Corp. Elk Grove Village, IL	1975	–	1965
Pettiford and Pettiford Contractors Inc. West Orange, NJ	1980	–	1965

	Joined NRCA	No. of Generations	Founded
Scott Roofing Co., Inc. Oceanside, CA	1978	–	1965
Sioux Roofing Co., Inc. Rock Rapids, IA	1980	–	1965
Stahl Inc. Brommfield, CO	1985	–	1965
Terry's Roofing and Sheet Metal Inc. Bossier City, LA	1980	1	1965
Weather-Reps Inc. Waltham, MA	1980	–	1965
Whitley Roofing Co. Richmond, VA	1966	–	1965
Willards Wholesale Roofing Co., Inc. Oklahoma City, OK	1986	3	1965

1966–1986

	Joined NRCA	No. of Generations	Founded
Alva Roofing Co., Inc. Alva, OK	1982	1	1966
Mike Barrow Roofing Co., Inc. North Little Rock, AR	1981	1	1966
Bradco Supply Corp. Avenel, NJ	1972	1	1966
Esary Roofing and Siding Co., Inc. Burlington, WA	1984	–	1966
Hamilton Roofing Co. Artesia, NM	1986	–	1966
J and B Roofing Corp. Cohoes, NY	1974	–	1966
Marathon Roofing Co., Inc. Marathon, FL	1983	–	1966
McClung Roofing Co., Inc. Fort Worth, TX	1983	1	1966
Missoula Sheet Metal Missoula, MT	1983	–	1966
Phillips Fibers Corp. Greenville, SC	1982	–	1966
C. R. Pidock Inc. Jacksonville, FL	1979	1	1966
Reeves Roofing Equipment Co., Inc. San Antonio, TX	1971	3	1966
Santo J. Ruisi Roofing Co., Inc. Deer Park, NY	1980	1	1966
Skyline Roofing Co., Inc. Manchester, NH	1971	–	1966
Super 'K' Industries Flint, MI	1977	3	1966
Western Building Materials Co. Fresno, CA	1974	–	1966
Adams Enterprises Inc. San Carlos, CA	1973	–	1967
Allen Roofing Co., Inc. Roswell, NM	1974	2	1967
Amrco Corp. Sterling Heights, MI	1970	–	1967
Arapahoe Roofing and Sheet Metal Inc. Broomfield, CO	1978	1	1967
Charbonneau Contracting Corp. Round Lake, NY	1976	–	1967

	Joined NRCA	No. of Generations	Founded
Charter Roofing Inc. Houston, TX	1981	2	1967
Dalton Roofing Co. Cincinnati, OH	1979	3	1967
Fabral-Alcan Building Products Lancaster, PA	1985	–	1967
Glas-Shield Roof Systems Portland, OR	1984	–	1967
Grieme Roofing Co., Inc. Pinehurst, NC	1985	1	1967
Havasu Roofing Inc. Lake Havasu City, AZ	1984	2	1967
Hillsdale Industries Inc. Knoxville, TN	1977	–	1967
Hoechst Fibers Industries Spartanburg, SC	1983	–	1967
Marland Roofing and Siding Co., Inc. Glen Mills, PA	1978	2	1967
Robert D. Monahan Associates Inc. Hudson Falls, NY	1978	1	1967
Republic Roofing Co. Kirkwood, MO	1974	–	1967
Roof Management Materials Co. Sun Valley, CA	1985	1	1967
States Roofing and Metal Co., Inc. Montgomery, AL	1974	1	1967
Tri-City Roofing Co., Inc. Kent, WA	1983	–	1967
United McGill Corp. Columbus, OH	1985	2	1967

	Joined NRCA	No. of Generations	Founded
Van Doorn Roofing Inc. Elk Grove, IL	1979	1	1967
Van Doorn Roofing Inc. Rolling Meadows, IL	1986	1	1967
Waters Brothers Contractors Rocky Mount, NC	1984	1	1967
Acme Roofing and Sheet Metal Co., Inc. Dothan, AL	1979	2	1968
Aetna Roofing Corp. Trenton, NJ	1981	1	1968
All Star Sheet Metal and Roofing Inc. Amarillo, TX	1983	1	1968
Apache Building Products Linden, NJ	1968	–	1968
Atlas-Apex Roofing Ltd. Ontario, Canada	1985	–	1968
B and L Sheet Metal and Roofing Inc. Bloomington, IN	1970	2	1968
Bornor Restoration Inc. Lansing, MI	1986	2	1968
Bruttell Roofing Inc. Oak Park, MI	1982	1	1968
Canfield Roofing Inc. Englewood, CO	1981	–	1968
Central States Roofing Co. Ames, IA	1971	–	1968
Chapman Brothers Roofing Co. East Haven, CT	1984	–	1968

	Joined NRCA	No. of Generations	Founded
Circle Roofing Co. Bellwood, IL	1968	–	1968
CEI Industries Howell, MI	1974	2	1968
Dennies Contracting Co., Inc. Memphis, TN	1959	1	1968
Diekevers Roofing Inc. Grand Rapids, MI	1986	2	1968
Florida Horizons Inc. Venice, FL	1970	1	1968
William H. Gallagher Inc. Crugers, NY	1985	3	1968
J. Giarnella and Son Inc. Bronx, NY	1986	–	1968
L. C. Heath Roofing Inc. Newport News, VA	1980	1	1968
Interstate Roofing and Waterproofing Inc. Onalaska, WI	1979	1	1968
Kustom Sheet Metal and Roofing Co., Inc. Akron, OH	1973	3	1968
Curtis W. Landrum Inc. Richmond, VA	1979	1	1968
Newt and Butch Roofing and Sheet Metal Clovis, NM	1982	–	1968
Northern Industrial Maintenance Bethel Park, PA	1981	–	1968
Oak Roofing and Sheet Metal Works Co., Inc. Woburn, MA	1977	2	1968

	Joined NRCA	No. of Generations	Founded
Owen Roofing of Hayward Hayward, CA	1978	–	1968
Roberts Roofing Co., Inc. Tucson, AZ	1985	1	1968
Roberts-McNutt Inc. North Little Rock, AR	1982	–	1968
Roofing and Contractors Equipment Co. Bensenville, IL	1970	–	1968
Runnebohm Construction Inc. Shelbyville, IN	1982	1	1968
Ed Shook Jr. Contractors Pittsburgh, PA	1986	–	1968
Skinner Roofing and Insulation Inc. Grand Forks, ND	1981	2	1968
Southeast Roofing and Sheet Metal Inc. Jacksonville, FL	1979	–	1968
Southern Distributors Corp. Garden Grove, CA	1981	–	1968
Tabor Inc. Dunbar, WV	1985	1	1968
Universal Roofers Inc. Phoenix, AZ	1970	–	1968
Brian R. White Co., Inc. Ukiah, CA	1980	–	1968
Wigand Roofing Co. Ardsley, PA	1974	1	1968
Able Roofing and Siding Co., Inc. CLayton, NJ	1984	–	1969

	Joined NRCA	No. of Generations	Founded
Arrow Roofing and Supply Inc. Wyoming, MI	1976	–	1969
Bean and Mallow Inc. Stafford, VA	1976	–	1969
Ron Case Roofing Inc. Salt Lake City, UT	1985	2	1969
Central Alabama Metal and Roofing Co. Montgomery, AL	1984	2	1969
Conklin Co., Inc. Minneapolis, MN	1979	1	1969
CEI Industries of Texas Inc. Dallas, TX	1982	2	1969
Danker Roofing Inc. Manhattan, KS	1980	1	1969
Doran Roofing Co., Inc. Fair Haven, VT	1980	2	1969
G. M. Ellis Industrial Maintenance Inc. Newtown, PA	1982	–	1969
Empro Corp. Minneapolis, MN	1985	1	1969
ESE Machines Exton, PA	1986	–	1969
Georgia State University Atlanta, GA	1984	–	1969
H. K. Griffith Inc. Newark, DE	1979	1	1969
J and J Contracting Eldorado, IL	1976	1	1969
Phil Johnson Sales Co., Inc. Comstock Park, MI	1973	–	1969

	Joined NRCA	No. of Generations	Founded
E. M. Martin Inc. Charlottesville, VA	1980	1	1969
Mid Hudson Pam Corp. Kingston, NY	1979	1	1969
Mueller-Potteiger Inc. York, PA	1969	2	1969
Rey-Crest Roofing and Waterproofing Co. Los Angeles, CA	1980	2	1969
Roofing Consultants Inc. Los Angeles, CA	1983	–	1969
Rowell Sheet Metal Inc. Foxworth, MS	1985	–	1969
Santa Barbara Roofing Inc. Santa Barbara, CA	1986	1	1969
Schena Roofing and Sheet Metal Co., Inc. Mount Clemens, MI	1971	1	1969
Weathercraft Co. of Lincoln Lincoln, NE	1970	2	1969
A and A Roofing Co., Inc. Dallas, TX	1980	2	1970
American Heating and Roofing Cody, WY	1976	–	1970
Apex Roofing Co. El Paso, TX	1982	–	1970
B and W Roofing Co. Riverdale, GA	1982	3	1970
Brown Industries Inc. Farmerville, LA	1979	–	1970
C and M Roofing Co., Inc. Pinehurst, MA	1978	–	1970

	Joined NRCA	No. of Generations	Founded
Cambric/Stetson Des Moines, IA	1980	–	1970
Chesterfield Roofing Inc. Midlothian, VA	1980	1	1970
Cloyd Corp. Louisville, KY	1971	–	1970
Diamond Roofing Co., Inc. Mattydale, NY	1976	2	1970
Russ Elliott Inc. Oakland, CA	1981	2	1970
Hannin Roofing and Sheet Metal Co., Inc. Paducah, KY	1971	–	1970
Heritage Wholesalers Inc. Malden, MA	1980	–	1970
Independent Roofing Cleveland, OH	1985	–	1970
Insulation Systems Inc. Collierville, TN	1981	–	1970
L. A. Kennedy Inc. Phillipsburg, NJ	1984	1	1970
L and M Co., Division of Nieto Roofing Port Chester, NY	1982	–	1970
Loadmaster Systems Inc. Dallas, TX	1981	2	1970
Lowndes Roofing and Sheet Metal Inc. Valdosta, GA	1974	2	1970
Patrick J. McKenna Roofing Inc. Pawtucket, RI	1982	2	1970

	Joined NRCA	No. of Generations	Founded
Modern Heating Cooling and Roofing Inc. Clarksville, TN	1986	–	1970
Morello Brothers Construction Inc. Medford, MA	1980	1	1970
P and C Roofing Inc. Wilmington, DE	1982	–	1970
Pacer Corp. Custer, SD	1983	–	1970
Pearce-Phelps Roofing Inc. Lexington, KY	1971	2	1970
Polymer Building Systems Inc. Riverside, CA	1977	–	1970
RCI Roofing Inc. Chicago, IL	1986	–	1970
R Mar Corp. Roofing and Materials San Jose, CA	1980	1	1970
Ristow Roofing Co., Inc. Kiel, WI	1982	2	1970
Schnabels Roofing Corp. Ronkonkoma, NY	1984	2	1970
Richard M. Smith Roofing Inc. Columbus, OH	1981	1	1970
Specialty Roofing Inc. Peoria, AZ	1979	2	1970
G. F. Sprague and Co., Inc. Braintree, MA	1982	–	1970
G. V. Sutton Inc. Dover, PA	1986	1	1970
Ann Arbor Roofing Ann Arbor, MI	1984	1	1971

	Joined NRCA	No. of Generations	Founded
Awaitam Roofs and Waterproofing Co. Lagos, Nigeria	1985	–	1971
Baker Roofing Inc. Travelers Rest, SC	1982	–	1971
Barry Roofing Co., Inc. Kinnelon, NJ	1984	–	1971
Batchelor Roofing Inc. Homestead, FL	1980	–	1971
Bennett and Brosseau Roofing Inc. Kankakee, IL	1984	–	1971
Bosnick Roofing Inc. Tacoma, WA	1977	1	1971
C and K Roofing Co. Miami, FL	1980	–	1971
C and S Roofing and Sheet Metal Co., Inc. Perth Amboy, NJ	1980	–	1971
Carolina Roofing Service Inc. Monroe, NC	1978	–	1971
Catana Roofing Co. Mount Clemens, MI	1979	2	1971
Coleman Roofing Inc. Chicago Heights, IL	1982	1	1971
Dodson Brothers Roofing Inc. Tulare, CA	1980	2	1971
First Property Management Corp. Chicago, IL	1981	–	1971
Foam Applications Inc. Huntsville, AL	1985	2	1971
H and R Roofing and Sheet Metal Inc. Sioux Falls, SD	1977	–	1971
Hoag Roofing Medford, OR	1984	–	1971
Intercoastal Maintenance Inc. Baltimore, MD	1985	–	1971
Jimco Products Livonia, MI	1983	1	1971
Martin E. Keller Roofing Co. Schenectady, NY	1972	1	1971
Kelly Energy Systems Inc. Waterbury, CT	1982	–	1971
Leigh's Roof Service Inc. McMinnville, OR	1981	–	1971
Mecklenburg Roofing Inc. Charlotte, NC	1974	–	1971
Resco Inc. Denver, CO	1973	–	1971
Richmond Roofing Co., Inc. Ashland, VA	1980	–	1971
Roberts' Maintenance Dundas Ltd. Ontario, Canada	1979	1	1971
Roof Decks of Puerto Rico Inc. San Juan, PR	1975	–	1971
Rue and Butler AIA Tacoma, WA	1985	–	1971
C. Walter Smith Roofing Contractors Everett, WA	1977	–	1971
Southern Colorado Roofing Co. Pueblo, CO	1984	3	1971

	Joined NRCA	No. of Generations	Founded
Success Roofing Inc. Kent, WA	1978	1	1971
Swift Roofing Inc. Murray, KY	1984	–	1971
Troutman Roofing Co., Inc. Troutman, NC	1984	–	1971
Vern and Sons Roofing Contractors Wallace, MI	1985	2	1971
White Roofing Co., Inc. Eldridge, IA	1984	–	1971
Woodhull Roofing Ltd. Birmingham, England	1981	2	1971
AR Contractors Inc. Fort Worth, TX	1986	–	1972
AC Products Inc. Placentia, CA	1985	–	1972
Cardinal Construction Co. Wooster, OH	1986	–	1972
Cedars West Roofing Inc. Boise, ID	1982	–	1972
Channel Islands Roofing Inc. Oxnard, CA	1984	–	1972
Collins Roofing Inc. Orem, UT	1980	1	1972
Colony Roofing Inc. New London, CT	1984	1	1972
Copper Sales Inc. Minneapolis, MN	1985	–	1972
Cross Roofing Inc. Meridian, MS	1977	–	1972

	Joined NRCA	No. of Generations	Founded
Design and Engineering Systems Inc. Fremont, CA	1986	–	1972
Dynamic Roofing Co., Inc. Rancho Cordova, CA	1985	–	1972
East Coast Roofing and Metals Inc. Greenville, NC	1981	1	1972
Len Eckstein Roofing Sunman, IN	1978	1	1972
Exeter Roofing and Sheet Metal Co., Ltd. Ontario, Canada	1985	–	1972
A. M. Fogarty and Associates Inc. Hingham, MA	1985	2	1972
Giles Sales Co., Inc. Scottdale, GA	1979	1	1972
Helms Roofing Co., of New Mexico Hobbs, NM	1979	–	1972
Hester Roofing Sacramento, CA	1972	–	1972
Hilo-Pacific Roofing Co., Inc. Hilo, HI	1978	2	1972
Hollis Roofing Inc. Columbus, MS	1984	–	1972
Independent Roofing Consultants Inc. Costa Mesa, CA	1984	–	1972
KR Roofing Co. Portland, OR	1985	1	1972
C. R. King Co. Linden, NJ	1977	–	1972

	Joined NRCA	No. of Generations	Founded
Lakeside Roofing Co. Granite City, IL	1976	–	1972
Merriam Roofing and Sheet Metal Co., Inc. Adamsville, OH	1980	2	1972
Monarch Building Supply Honolulu, HI	1985	–	1972
MCA Roofing Butte, MT	1984	–	1972
Navasota Roofing and Sheet Metal Co., Inc. Navasota, TX	1978	1	1972
Nightingale Roofing Inc. Peterborough, NH	1978	–	1972
Northern Ohio Roofing Inc. Elyria, OH	1979	–	1972
Northern Roofing (1972) Ltd. Ontario, Canada	1980	–	1972
Palmetto State Roofing and Sheet Metal Co., Inc. Greenville, SC	1979	2	1972
Pawcatuck Roofing Co., Inc. Pawcatuck, CT	1978	2	1972
Roof Mechanics Inc. Wichita, KS	1974	–	1972
Roofs Ltd. of Virginia Alexandria, VA	1985	2	1972
Paul J. Rys Roofing Inc. Philadelphia, PA	1982	1	1972
SQI Inc. Seattle, WA	1977	–	1972

	Joined NRCA	No. of Generations	Founded
Sennott Roofing Co., Inc. Winthrop, MA	1981	–	1972
D. Seymore Contractor Inc. Dysart, PA	1981	–	1972
Sharp Roofing Sales Inc. Rogue River, OR	1984	1	1972
Smith Roofing Plattsburgh, NY	1985	–	1972
Vanderbroek Roofing Inc. Grand Rapids, MI	1975	–	1972
Vanguard Organization Inc. Poughkeepsie, NY	1981	1	1972
Vohnoutka Roofing and Sheet Metal Co., Inc. Eden Prairie, MN	1973	1	1972
Waldrep and Sons Roofing Inc. Olen Davie, FL	1980	2	1972
G.N. Willard Roofing Co. Knoxville, TN	1974	1	1972
All Seasons Roofing Lafayette, IN	1985	1	1973
Atlantic Asphalt and Equipment Co., Inc. Revere, MA	1983	1	1973
Barrier Roofing and Sheet Metal Co., Inc. Canton, OH	1975	–	1973
A. Bartlett Roofing Inc. Gainesville, FL	1985	–	1973
Blue's Roofing Co. San Jose, CA	1974	2	1973

	Joined NRCA	No. of Generations	Founded
Michael J. Bollinger Co. Baltimore, MD	1984	–	1973
Bollinger Roofing Co., Inc. Baltimore, MD	1973	2	1973
Brown's Roofing Millville, NJ	1978	–	1973
Buttweiler's Do-All Inc. Alexandria, MN	1978	–	1973
Cheesbro Roofing Inc. Ormond Beach, FL	1979	1	1973
Estimation Inc. Linthicum Heights, MD	1986	–	1973
General Roofing and Heating Co., Inc. Wellsville, NY	1981	–	1973
Granite City Roofing Inc. St. Cloud, MN	1978	–	1973
Gulfside Supply Inc. Tampa, FL	1983	1	1973
Hayden Enterprises Inc. Canyon Country, CA	1986	–	1973
A. W. Hepp and Co., Inc. Little Rock, AR	1984	1	1973
Independent Roofing Inspectors West Chester, PA	1983	2	1973
G. L. Kautz Inc. Lancaster, PA	1981	1	1973
Kelco Roofing Co., Inc. Atlanta, GA	1974	1	1973
Life/Roofing Division Chamberlain, SD	1982	1	1973

	Joined NRCA	No. of Generations	Founded
Liquid Asphalt Systems Inc. Kansas City, MO	1973	–	1973
Logan Roofing Co. Mullica Hill, NJ	1978	–	1973
Maco Roof Systems Inc. Wheeling, IL	1977	–	1973
Mike McAdams Roofing Co., Inc. Pascagoula, MS	1980	1	1973
Meier Roofing Co. Elk Grove, IL	1982	1	1973
Midwest Roofing Co. Houston, TX	1984	–	1973
Mike's Roofing Service Inc. Van Nuys, CA	1983	–	1973
Molnar Group Inc. Solon, OH	1985	–	1973
Operculum Corp. Whitefish, MT	1986	–	1973
P and P Roofing Inc. Victoria, TX	1984	–	1973
Pieros Construction Co., Inc. Somerville, NJ	1983	–	1973
Potter and Son Roofing Inc. Waco, TX	1984	2	1973
Quality Roofing Co., Inc. Passaic, NJ	1979	1	1973
Quality Roofing Co., Inc. Manassas, VA	1979	–	1973
RR Roofing and Construction Co. Chicago, IL	1980	–	1973

	Joined NRCA	No. of Generations	Founded
Roof Maintenance Systems Farmingdale, NJ	1982	–	1973
Roofing Enterprises Hingham, MA	1979	1	1973
Rudderow Roofing of Georgia Inc. Marietta, GA	1986	1	1973
Runnion Equipment Co. Lyons, IL	1982	1	1973
Southern Reserve Roofing Co. Atlanta, GA	1975	2	1973
Sta-Dri Co., Inc. Dallas, TX	1985	2	1973
SRD Inc. Pleasant Grove, AL	1983	–	1973
Target Roofing Inc. Devils Lake, ND	1981	–	1973
TAM Construction Co., Inc. Olive Branch, MS	1974	–	1973
Upstate Roofing Inc. Rochester, NY	1977	1	1973
Venture Roofing Co., Inc. Flagstaff, AZ	1984	1	1973
Washington Roofing and Sheet Metal Co., Inc. Huntington, WV	1979	2	1973
Western Roofing Co., Ltd. Alberta, Canada	1976	–	1973
Wolfe Roofing and Sheet Metal Inc. Reading, PA	1979	2	1973
Advance Group Enterprises Inc. Washington, DC	1984	–	1974

	Joined NRCA	No. of Generations	Founded
Associated Roofers Inc. Chicago, IL	1977	–	1974
Brinkmann Roofing and Sheet Metal Inc. Webster, TX	1982	1	1974
Cellular Product Services Inc. Colorado Springs, CO	1983	–	1974
Coates Roofing Co., Inc. Choctaw, OK	1983	2	1974
Columbus Roofing Inc. Columbus, GA	1983	2	1974
DW Roofing Midland, TX	1986	1	1974
Davis Roofing Corp. Chandler, AZ	1978	–	1974
The Denchfield Corp. Washington, DC	1981	1	1974
Edwards Christensen Roofing (1983) Ltd. Alberta, Canada	1979	–	1974
Elite Roofing and Sheet Metal Inc. Cudahy, WI	1984	–	1974
Z. Ellis Roofing Co., Inc. Kenner, LA	1977	1	1974
Ferber and Osteen Roofing Co., Inc. Gainesville, FL	1980	–	1974
Harrell Roofing Inc. Tallahassee, FL	1985	–	1974
Jottan Co., Inc. Evans City, PA	1977	–	1974

	Joined NRCA	No. of Generations	Founded
Latham Roofing Inc. Dallas, TX	1978	2	1974
Lawmaster Inc. Elkhart, IN	1975	1	1974
LaFleche Roofing Ltd. Ontario, Canada	1985	1	1974
Stark Lewis Co. Anchorage, AK	1969	1	1974
Lorenz and Associates St. Louis, MO	1983	1	1974
New England Weatherproofing Corp. Winthrop, MA	1978	–	1974
NIKO Contracting Co. Pittsburgh, PA	1985	–	1974
Peterman Roofing Seattle, WA	1981	–	1974
Porter Roofing Co., Inc. Belton, MO	1978	2	1974
Primo Roofing Co. Huntington Beach, CA	1980	–	1974
R and R Roofing and Sheet Metal Inc. Ironton, OH	1981	2	1974
Rackley Roofing Co., Inc. Carthage, TN	1975	1	1974
Rain-Master Inc. Fort Wayne, IN	1986	1	1974
Ridgeworth Roofing Co., Inc. Palos Heights, IL	1976	–	1974
Roofers Mart of Oregon Inc. Portland, OR	1978	–	1974

	Joined NRCA	No. of Generations	Founded
Roofs Inc. Lyons, IL	1978	2	1974
Service Roofing and Sheet Metal Co. Greenville, NC	1979	–	1974
Skinner Roofing and Sheet Metal Colorado Springs, CO	1976	–	1974
Specialty Roofing Co., Inc. Union City, GA	1981	2	1974
Standard Home and Industry Inc. Oklahoma City, OK	1982	–	1974
Standard Spray Systems Inc. Raleigh, NC	1986	–	1974
Superior Roofing Contractors Inc. Montgomery, AL	1986	–	1974
Texas Fifth Wall Roofing Systems Inc. Austin, TX	1980	1	1974
Waterproofing Systems Miami Miami, FL	1983	–	1974
All Roofing Co. Warren, MI	1982	–	1975
All South Subcontractors Inc. Nashville, TN	1975	–	1975
B and M Sheet Metal Roofing Inc. Wichita Falls, TX	1976	1	1975
Boss Insulation and Roofing Co. Lewisburg, PA	1985	1	1975
R. A. Bousquet Inc. Plainfield, NJ	1983	1	1975
Caesar's Roofing Inc. Wakefield, MA	1981	–	1975

	Joined NRCA	No. of Generations	Founded
Commercial Roofing Inc. Plover, WI	1981	–	1975
Consolidated Enterprises Inc. Anchorage, AK	1982	–	1975
Cornerstone Roofing Co. Albuquerque, NM	1984	–	1975
Culbertson Co. of Va. Inc. Manassas, VA	1982	–	1975
M. J. Dalsin Co., Inc. West Fargo, ND	1978	1	1975
Dewmor Roofing Corp. East Syracuse, NY	1980	1	1975
Edwards Roofing Thayer, IL	1986	–	1975
Florida Roofing and Sheet Metal Co. Fort Myers, FL	1975	–	1975
G and E Roofing Co., Inc. Augusta, ME	1982	–	1975
Gorman Roofing Inc. Cumberland, RI	1985	1	1975
Gracie Roofing Inc. Burlington, VT	1980	1	1975
Jim Griffin Roofing Co., Inc. Charlotte, NC	1984	2	1975
Hammer Roofing Inc. Broomfield, CO	1984	–	1975
Higgs Inc. Parkersburg, WV	1983	1	1975
Maddox Roofing Construction Hobson, MT	1978	3	1975

	Joined NRCA	No. of Generations	Founded
Manson Roofing Inc. Bradenton, FL	1986	1	1975
Marshall Roofing Inc. Fort Lauderdale, FL	1982	–	1975
McGaughey Building Products Milan, TN	1984	2	1975
McLaughlin Roofing Corp. Elmira, NY	1982	1	1975
Murton Roofing Corp. Miami, FL	1977	2	1975
Neidig Roofing Co. Harrisburg, PA	1977	–	1975
NRG Barriers Inc. Sanford, ME	1980	–	1975
Raintree Industries Inc. Emigsville, PA	1978	–	1975
William Samuels Approved Roofing Inc. Bradenton, FL	1978	1	1975
Sarnafil Inc. Canton, MA	1981	–	1975
Ronald J. Seacord Ithaca, NY	1979	–	1975
Spartan Roofing Co., Inc. Lansing, MI	1978	1	1975
Suncoast Roofers Supply Inc. St. Petersburg, FL	1981	–	1975
Tech Roofing Service Inc. Springfield, MA	1979	–	1975
Tuscano-Maher Roofing Saltsburg, PA	1985	–	1975

	Joined NRCA	No. of Generations	Founded
Vacuum Engineering Corp. Milwaukee, WI	1978	–	1975
VMS Diversified Inc. Redmond, WA	1984	1	1975
Weathertite Systems Inc. Grand Island, NE	1980	–	1975
Westover Products Inc. Greensboro, NC	1981	2	1975
E. J. Wilder and Sons Inc. Bradley, IL	1978	1	1975
Wyoming Roof Service Inc. Casper, WY	1980	–	1975
Able Sheet Metal and Roofing Houston, TX	1984	–	1976
B and R Roofing and Sheet Metal Co., Inc. Hays, KS	1980	–	1976
Baron and Wheeler Inc. Pittsburgh, PA	1980	–	1976
Black Roofing Inc. Boulder, CO	1980	–	1976
Bruton-Gomez-Easley Inc. Corpus Christi, TX	1980	–	1976
Builders Custom Roofing Inc. Kenmore, WA	1984	–	1976
Building Technicians Corp. Geneva, OH	1979	–	1976
BAB Roofing Co., Inc. Flint, MI	1969	–	1976
Wayne Cain and Sons Roofing and Sheet Metal Champaign, IL	1978	2	1976
Campeon Roofing and Waterproofing Inc. Cincinnati, OH	1984	–	1976
Cascade Empire Corp. Portland, OR	1984	–	1976
W. J. Collum Roofing Co., Inc. Tempe, AZ	1979	2	1976
DJ Roofing Inc. Santa Ana, CA	1984	–	1976
Ducks Roofing Co., Inc. Newport News, VA	1983	–	1976
Dutchglas Roofing Systems Laguna Hills, CA	1976	–	1976
Energy Shield Inc. Pontiac, MI	1982	2	1976
Flex-Shield Corp. Gilbert, AZ	1986	2	1976
Fort Roofing of Columbia Inc. Columbia, SC	1976	–	1976
G and S Roofing Inc. Minneapolis, MN	1982	1	1976
Gage Contractors South Houston, TX	1986	–	1976
Greensboro Roofing Co., Inc. Greensboro, NC	1981	2	1976
Greg's Roofing Inc. Ormond Beach, FL	1985	–	1976
HRDC Construction Minneapolis, MN	1984	1	1976
John Henry Roofing Inc. Brighton, MA	1981	–	1976

	Joined NRCA	No. of Generations	Founded
K and L Roofing Co., Inc. Phoenix, AZ	1980	–	1976
King Asphalt Products Co., Inc. Douglasville, GA	1981	–	1976
LaFerney Inc. Kingsport, TN	1985	1	1976
Lydick Roofing of Wichita Falls Inc. Wichita Falls, TX	1977	–	1976
M and M Roofing Inc. South Holland, IL	1986	–	1976
Marton Roofing Inc. Houston, TX	1985	1	1976
Metal Building Components Inc. Houston, TX	1983	–	1976
Metropolitan Roofing Co., Inc. Eden, NC	1985	–	1976
J. L. Murray Co., Inc. West Seneca, NY	1983	1	1976
Nationwide Roofing and Sheet Metal Inc. Dayton, OH	1980	–	1976
North Roofing Co. Pontiac, MI	1976	–	1976
Northwestern Roofing Co., Inc. Meadville, PA	1969	–	1976
Phoenix Roofing Inc. Phoenix, AZ	1980	–	1976
Pleasant Valley Roofing Springtown, PA	1984	–	1976
Pocatello Roofing Pocatello, ID	1984	2	1976

	Joined NRCA	No. of Generations	Founded
Clinton Powell Roofing and Construction Cleveland, MS	1985	1	1976
Precision Roof Systems Inc. San Jose, CA	1981	–	1976
Preventative Maintenance Irving, TX	1984	–	1976
Progressive Roofing Inc. Buffalo, NY	1979	1	1976
Quality Roofing Contractors of Southeast Missouri Inc. Senath, MO	1981	–	1976
RSI Industries Inc. Cape Coral, FL	1984	2	1976
Rayco Roof Service Inc. Chantilly, VA	1980	2	1976
Redd Roofing Co., Inc. Ogden, UT	1981	–	1976
Roewer Roofing McHenry, IL	1985	1	1976
Roof Builders Inc. El Paso, TX	1982	–	1976
Roof Right Roofing and Insulation Co. Posen, IL	1978	1	1976
Roofs Inc. Lexington, KY	1980	–	1976
RSI Wholesale Inc. Grand Rapids, MI	1983	–	1976
Servcon Marketing Inc. Quebec, Canada	1982	–	1976

	Joined NRCA	No. of Generations	Founded
Southeast Building Maintenance Tech Inc. Doraville, GA	1981	2	1976
A. Robert St. Hilaire Inc. Auburn, ME	1976	–	1976
Thomson Roofing and Metal Co. Thomson, GA	1983	2	1976
Tri Spec Systems Inc. Creve Coeur, MO	1980	1	1976
Valentine Roofing Co., Inc. Houston, TX	1982	–	1976
Van Winkle Roofing Inc. Roswell, NM	1976	2	1976
Wayne's Roofing Inc. Sumner, WA	1980	2	1976
Weathercraft Co. of North Platte North Platte, NE	1980	–	1976
Able Roofing Co., Inc. Bloomington, IN	1984	–	1977
American Hydrotech Chicago, IL	1983	–	1977
Associates Roofing Hollywood, FL	1986	–	1977
Aulson Roofing Inc. Peabody, MA	1985	1	1977
Beaver Enterprises Inc. Atlanta, GA	1980	2	1977
Carpenter Insulation and Coatings Co. Dallas, TX	1982	–	1977
City Roofing Co., Inc. Memphis, TN	1979	1	1977

	Joined NRCA	No. of Generations	Founded
Cleo's Roofing Inc. Tacoma, WA	1983	2	1977
Consolidated Roofing Inc. Denver, CO	1983	–	1977
CBC Enterprises Inc. Denver, CO	1978	–	1977
D and D Roofing Inc. Salisbury, MD	1981	1	1977
Damiano Roofing Co. Elk Grove Village, IL	1984	–	1977
J. Detorie Roofing Co., Inc. Baltimore, MD	1980	3	1977
Dickson Roofing Inc. Tulsa, OK	1982	–	1977
Divins Roofing Co. Maitland, FL	1980	–	1977
Dixon Roofing Inc. Canal Winchester, OH	1985	2	1977
Eastern Roofing Corp. Norfolk, VA	1978	2	1977
Evergreen Roofing of Oregon Inc. Portland, OR	1984	2	1977
Florida State Roofing Inc. North Miami Beach, FL	1984	1	1977
Fredericks Roofing Pendleton, IN	1985	2	1977
Fulton Roofing and Construction Co., Inc. Mobile, AL	1983	2	1977
Georgia Building Maintenance Atlanta, GA	1982	–	1977

	Joined NRCA	No. of Generations	Founded
Grainger Associates Inc. Flint, MI	1986	–	1977
Grainger Building Services Ltd. North Ireland	1982	2	1977
Imperial Roof Systems West Union, IA	1986	1	1977
Industrial Roofing Inc. Anchorage, AK	1984	–	1977
Kiepura-Stewart Roofing Inc. Chicago Heights, IL	1979	1	1977
Mak-Lyn Inc. Brackenridge, PA	1979	–	1977
McRae Roofing and Siding Corp. Asheboro, NC	1983	2	1977
R. E. Miller Co. Dallas, TX	1979	–	1977
L. W. Miller Roofing Inc. Washington, PA	1986	2	1977
Misnad Bin Sa'ad Al-Misnad Est. Qatar, Arabia	1984	–	1977
Tom Moore Roofing and Construction Co. Mount Prospect, IL	1981	1	1977
National Roofing Co., Inc. Albuquerque, NM	1985	–	1977
NPS Corp. Perryville, MO	1984	–	1977
On Top Roofing Inc. Grandview, MO	1978	–	1977
Portals Plus Inc. Bensenville, IL	1985	–	1977

	Joined NRCA	No. of Generations	Founded
Michael A. Prate Inc. Wauconda, IL	1985	–	1977
Ratliff Roofing and Supply Inc. Tucson, AZ	1986	2	1977
J. L. Robinson Roofing Co. Pittsburgh, PA	1982	1	1977
Roof Design Specialists Inc. Santa Ana, CA	1980	1	1977
Roof-Top Engineers Inc. Phoenix, AZ	1978	–	1977
Roofing and Sheet Metal Inc. Huntsville, AL	1986	2	1977
Sandra Corp. North Prairie, WI	1983	–	1977
J. E. Shomo and Son Inc. Somers Point, NJ	1980	–	1977
Slavik and Butcher Construction Co. Auburn Heights, MI	1984	–	1977
Sourbeck Roofing Inc. St. Petersburg, FL	1983	2	1977
Southwest Roofing Co. Farmington, NM	1984	1	1977
Super Utomo Corp. Jakarta, Indonesia	1986	–	1977
Travis Roofing Supply Inc. Austin, TX	1979	–	1977
Triangle Fastener Corp. Pittsburgh, PA	1986	–	1977
US Roofing Inc. Peabody, MA	1979	–	1977

	Joined NRCA	No. of Generations	Founded
F. R. Ulrich and Sons Inc. Baltimore, MD	1978	–	1977
Williams Roofing Co., Inc. Castroville, CA	1982	1	1977
A and M Roofing and Sheet Metal Co., Inc. Middleton, MA	1980	1	1978
A-Alltate Roofing Denver, CO	1984	1	1978
Advanced Roofing and Construction Inc. Newark, NJ	1986	3	1978
Affiliated Roofing Enterprises Santa Rosa, CA	1981	–	1978
Amerson Roofing and Siding Co. Monroeville, AL	1984	1	1978
Anthony Roofing Ltd. Aurora, IL	1981	1	1978
Arrow Roofing Inc. Phoenix, AZ	1982	1	1978
ABC Sandron Corp. Plano, TX	1984	1	1978
B and A Roofing Co., Inc. Manchester, GA	1981	1	1978
Best Roofing Fort Lauderdale, FL	1980	–	1978
P. B. Brodak Roofing and Sheet Metal Co., Inc. Wixom, MI	1985	–	1978
Tom Brown Contracting Inc. Pittsburgh, PA	1983	–	1978

	Joined NRCA	No. of Generations	Founded
CMS Roofing Inc. Fort Wayne, IN	1963	–	1978
Capitol Roofing Service Sandy, UT	1986	–	1978
Ciro Roofing Products USA Inc. Custer, WA	1981	1	1978
Columbia ABECE Inc. Vancouver, WA	1983	–	1978
Columbiana Roofing Inc. Lowellville, OH	1985	1	1978
Comdustrial Roofing Contractors Inc. Hatfield, PA	1981	–	1978
Craftmasters Inc. Decatur, IL	1980	2	1978
Dehling Voigt Inc. Newburg, WI	1980	1	1978
ERC Roofing Corp. Seneca Falls, NY	1986	1	1978
Exterior Maintenance Contracting Inc. Charlotte, NC	1981	1	1978
Ford Construction Superior, WI	1986	–	1978
Frost Construction Dayton, OH	1985	–	1978
Gale Roofing Co., Inc. Hart, MI	1983	–	1978
Genstar Roofing Products Co. Irving, TX	1966	–	1978
Goerschler Roofing Co., Inc. Schaumburg, IL	1984	1	1978

	Joined NRCA	No. of Generations	Founded
Great Basin Roofing Inc. Salt Lake City, UT	1984	–	1978
Grosjean Roofing and Sheet Metal Bossier City, LA	1984	3	1978
Guaranteed Roofing Co., Inc. Albuquerque, NM	1986	–	1978
Haeber Roofing Co. Corpus Christi, TX	1983	1	1978
Industrial Roofing and Sheet Metal Nashville, TN	1980	2	1978
Israel's Roofing Co., Inc. Sherman, TX	1985	–	1978
JGA Corp. Doraville, GA	1979	1	1978
Kandl Roofing Inc. Villa Park, IL	1986	–	1978
A. Kirby and Sons Inc. White Marsh, MD	1979	1	1978
Phil R. Klapp Inc. Mayfield, KY	1984	1	1978
L Quinta Roofing Inc. San Diego, CA	1985	2	1978
Lanier Construction Systems Inc. Griffin, GA	1982	–	1978
Lundin Roofing Co., Inc. Baton Rouge, LA	1982	1	1978
Mahan Roofing Co., and Sheet Metal Inc. Olive Branch, MS	1984	1	1978
Maine Roofing Service Waterville, ME	1982	–	1978

	Joined NRCA	No. of Generations	Founded
Maris and Son Inc. Hobart, IN	1980	1	1978
Martin Roofing Co., Inc. Greensboro, NC	1980	–	1978
Mid-Atlantic Roofing Inc. Charlotte, NC	1981	2	1978
Missouri-Illinois Roof Systems Inc. St. Louis, MO	1981	3	1978
Mt. Hood Roofing Co., Inc. Troutdale, OR	1983	2	1978
James Myers Co., Inc. Beltsville, MD	1979	1	1978
New Castle Roofing and Waterproofing New Castle, DE	1983	2	1978
New London Roofing Co., Inc. New London, CT	1982	–	1978
Ozark Roofing of Springfield Inc. Springfield, MO	1979	2	1978
PCM Enterprises Inc. West Palm Beach, FL	1984	–	1978
Palm Bay Roofing and Sheet Metal, Inc. Palm Bay, FL	1978	–	1978
Palmer Roofing Co. Pendleton, OR	1978	1	1978
Pennsylvania Roofing Systems Inc. Bakerstown, PA	1980	–	1978
Plymouth Foam Products Plymouth, WI	1980	2	1978

	Joined NRCA	No. of Generations	Founded
Porter Roofing Contractors Inc. McMinnville, TN	1979	–	1978
Powers Roofing and Sheet Metal Inc. Caribou, ME	1985	2	1978
Professional Urethanes Lacrosse, WI	1985	2	1978
Rockwell Roofing Inc. Leominster, MA	1972	–	1978
Roofers Mart Inc. Columbia, SC	1979	–	1978
Roofing Systems Inc. Loves Park, IL	1979	3	1978
Rowell Roofing Inc. Columbia, MS	1980	–	1978
RMAX Inc. Dallas, TX	1980	–	1978
Scott Roofing and Repair Inc. Opa-Locka, FL	1980	–	1978
Shen Valley Roofing Bridgewater, VA	1985	1	1978
Shieldco Inc. Carlisle, MA	1984	1	1978
Siplast Inc. Irving, TX	1980	–	1978
State Roofers Inc. Monroe, WA	1985	1	1978
Structural Research Inc. Madison, WI	1983	–	1978
Ray Sykes and Sons Inc. Absecon, NJ	1981	2	1978

	Joined NRCA	No. of Generations	Founded
Thermo Scan Engineering Inc. Littleton, CO	1986	–	1978
David A. Tucker Roofing Lake Park, FL	1980	–	1978
Universal Applicators Inc. Forest Lake, MN	1986	–	1978
Van Dyne and Sons' Roofing Sparks, NV	1979	2	1978
Wat Pro Inc. Kimberton, PA	1981	–	1978
York Roofing Inc. York, PA	1980	2	1978
ARTech Inc. Columbus, OH	1983	–	1979
Accurate Roofing Sunrise, FL	1986	–	1979
Action Roofing Torrance, CA	1985	1	1979
All Weather Exteriors Inc. Mishawaka, IN	1980	3	1979
Ameier Roofing Inc. Chicago Ridge, IL	1981	–	1979
American Western Manufacturing Mesa, AZ	1981	–	1979
B and H Urethane Systems Inc. Las Vegas, NV	1982	1	1979
Barra Corp., of America Inc. West Caldwell, NJ	1979	–	1979
Martin J. Bergen Inc. Philadelphia, PA	1981	1	1979

	Joined NRCA	No. of Generations	Founded
Blake Brothers Construction Co., Inc. Canyon, TX	1986	–	1979
Bloom Roofing Systems Inc. Ann Arbor, MI	1985	–	1979
Bredemeyer Roofing Co., Inc. Ft. Wayne, IN	1985	1	1979
Browne and Merry Construction Co., Inc. Washington, DC	1983	–	1979
Robert D. Byrd and Associates Big Bear Lake, CA	1983	–	1979
Cactus State Roofers Phoenix, AZ	1981	–	1979
Cam-Field Inc. Berkley, MI	1985	–	1979
Castro Roofing Inc. Richardson, TX	1983	1	1979
Clerkin Roofing Colorado City, CO	1982	1	1979
Commercial Roofing and Waterproofing Billings, MT	1985	1	1979
Diamond Roofing Dodge City, KS	1979	–	1979
Diamond Roofing Co., Inc. Atlanta, GA	1980	1	1979
Dumar Inc. Somerville, NJ	1981	–	1979
Duro-Last Roofing Inc. Saginaw, MI	1982	2	1979

	Joined NRCA	No. of Generations	Founded
Eastern Roofing and Sheet Metal Inc. Douglassville, PA	1980	–	1979
Everett Roofing Inc. Baltimore, MD	1980	2	1979
H. B. Fishman and Co., Inc. South Windsor, CT	1971	2	1979
Futura Coatings Inc. Hazelwood, MO	1986	2	1979
Galco Building Products Anchorage, AK	1982	–	1979
Grand County Roofing Inc. Winter Park, CO	1986	–	1979
Halperin Management Associates Inc. Riviera Beach, FL	1983	1	1979
David F. Heath Inc. Hampton, VA	1983	1	1979
Henris Roofing and Supply of Klamath Falls, OR	1985	1	1979
Bob Hilson and Co., Inc. Miami, FL	1979	1	1979
HR General Maintenance Corp. Washington, DC	1984	1	1979
Insulated Building Sytems Inc. Sterling, VA	1983	–	1979
Interstate Coatings Inc. Wilson, NC	1982	–	1979
Kare Roofing Inc. Crystal Lake, IL	1985	–	1979
M. J. Kelleher and Associates Inc. Pleasanton, CA	1984	1	1979

	Joined NRCA	No. of Generations	Founded
Kevin and Sons Roofing Corp. Evergreen Park, IL	1982	1	1979
Kiker Corp. Mobile, AL	1980	1	1979
Lenco Roofing and Insulation Inc. Carmichael, CA	1984	–	1979
The Linc Corp. Manchester, CT	1982	–	1979
The J. B. Lovell Corp. Alpharetta, GA	1980	–	1979
McKenzie and Cross Inc. Fort Worth, TX	1984	–	1979
MJC Corp. Cheverly, MD	1981	1	1979
National Roofing Supplies Patchogue, NY	1986	–	1979
George Neel and Associates Inc. Hattiesburg, MS	1986	–	1979
Olympic Fasteners Agawam, MA	1982	–	1979
Preservation Specialties Inc. Arvada, CO	1983	–	1979
Radco/Synergetics Tempe, AZ	1983	–	1979
Reliable Guttering Roofing Service Inc. Lexington, KY	1983	–	1979
Rich Roofing Corp. Manchester, CT	1981	2	1979
Rocky Mountain Roofing Co. Albuquerque, NM	1984	–	1979

	Joined NRCA	No. of Generations	Founded
Roof-Tek Inc. Marshville, NC	1982	–	1979
Roofing Specialists Inc. Englewood, CO	1981	–	1979
Rounds Roofing Lancaster, CA	1985	–	1979
S and S Roof Maintenance Inc. Buffalo Grove, IL	1982	–	1979
SCF Decks Inc. Baltimore, MD	1982	–	1979
Scholten Roofing Service Co. Mission Viejo, CA	1980	–	1979
Slatile of Valparaiso Inc. Valparaiso, IN	1986	–	1979
The Sobeck Corp. Wyoming, PA	1979	2	1979
South Central Roofing and Sheet Metal Inc. Hattiesburg, MS	1984	–	1979
Sta-Dry Roofing Inc. Union City, TN	1985	2	1979
Louis A. Stilloe Roofing and Siding Inc. Binghamton, NY	1982	1	1979
Sutter Roofing Co., of Florida Longwood, FL	1982	–	1979
SYenergy Methods Inc. Cranston, RI	1980	–	1979
Thornton and Son Sales Inc. Tacoma, WA	1985	4	1979
Tri S International Tel Aviv, Israel	1986	–	1979

	Joined NRCA	No. of Generations	Founded
Tri-Ply Inc. Madison Heights, MI	1984	–	1979
Troco Roofing Inc. Huntington, IN	1986	2	1979
F. W. Walton Roofing Co. Texas City, TX	1979	1	1979
Weathercraft Co., Inc. Gering, NE	1981	–	1979
WeatherGard Roofing Systems Inc. Memphis, TN	1981	–	1979
West Roofing Supply Co., Inc. Louisville, KY	1986	1	1979
Whatco Roofing and Sheet Metal Inc. Beaverton, OR	1982	2	1979
WGM Roofing Systems Inc. Wauconda, IL	1985	1	1979
York Manufacturing Inc. Sanford, ME	1979	1	1979
ARC Roof Corp. Oakland, CA	1981	–	1980
A-B Roofing Co., Inc. Lexington, KY	1981	–	1980
Abingdon Roofing Co., Inc. Abingdon, VA	1982	–	1980
American Roofing Corp. Chicago, IL	1982	–	1980
Anderson and Shah Roofing Inc. Joliet, IL	1981	–	1980

	Joined NRCA	No. of Generations	Founded
Apex Roofing Co., Inc. Burr Ridge, IL	1985	1	1980
Arkra Inc. Lakewood, OH	1986	2	1980
Atlas Roof Systems Inc. Bossier City, LA	1981	1	1980
Bailey Enterprises Inc. Canadys, SC	1984	–	1980
Baker Roofing Co. Norfolk, VA	1984	3	1980
W. J. Butzen Roofing and Sheet Metal Inc. Hales Corners, WI	1980	1	1980
C and S Roofing Dunnellon, FL	1980	–	1980
Cabell Sheet Metal and Roofing Inc. Ceredo, WV	1986	1	1980
Cairo and Sons Roofing Co., Inc. Chicago, IL	1981	1	1980
Champion Roofing Inc. North Plainfield, NJ	1984	–	1980
David A. Chenoweth Roofing Inc. Three Rivers, MI	1981	1	1980
Columbine International Placerville, CA	1983	–	1980
Cooley Roofing Systems Inc. Pawtucket, RI	1980	2	1980
Cox Roofing Co. Springfield, MO	1984	–	1980
Crews Roofing Inc. Frankfort, IN	1981	2	1980

	Joined NRCA	No. of Generations	Founded
Crown Roofing Inc. Norco, CA	1986	1	1980
Designed Roofing Systems Inc. Woonsocket, RI	1980	–	1980
Dews Roofing Inc. Crewe, VA	1980	2	1980
Eagle Roofing Systems Corp. Dallas, TX	1980	–	1980
Ench Roofing Supply West Orange, NJ	1984	–	1980
Envirospec Inc. Buffalo, NY	1982	1	1980
Firestone Building Products Co. Indianapolis, IN	1980	–	1980
Al Gilmore Inc. Mableton, GA	1983	–	1980
Guaina Corp., of America Hackensack, NJ	1986	–	1980
Charles Haller Enterprises Inc. Syracuse, NY	1985	–	1980
Hittle Roofing Systems Inc. Zanesville, OH	1985	2	1980
J and P Roofing Inc. Mansfield, IL	1983	–	1980
Kelley and Co., Inc. Ft.Wayne, IN	1985	–	1980
Kokem Products Inc. Portland, OR	1983	1	1980
T. W. Lakeside Roofing Inc. Bloomingdale, IL	1986	1	1980
Lawson Roofing Inc. Springdale, AR	1984	–	1980

	Joined NRCA	No. of Generations	Founded
Lester's Roofing and Sheet Metal Inc. San Antonio, TX	1984	–	1980
Long Life Roofing Wisconsin Dells, WI	1985	1	1980
Low Country Roofing Services Inc. Charleston, SC	1981	–	1980
Mackey Roofing Co. Tualatin, OR	1984	–	1980
McKee Roofing Inc. Richmond, VA	1983	–	1980
Metal-Era Inc. Waukesha, WI	1985	–	1980
Mid-State Roofing Inc. Dover, DE	1981	1	1980
Midlands Roofing and Flooring Inc. Columbia, SC	1984	–	1980
R. J. Myron Enterprises Inc. Ontario, Canada	1986	1	1980
National Roof Deck Contractors Association Chicago, IL	1986	–	1980
Northern Roofing Contractors Anchorage, AK	1984	–	1980
Pro Fastening Systems Arlington Heights, IL	1985	–	1980
Rafoth Furnace and Sheet Metal Works Dubuque, IA	1976	–	1980
The Roofers Inc. Wilmington, DE	1984	–	1980

	Joined NRCA	No. of Generations	Founded
Roofing Enterprises of South Carolina. Inc. Sumter, SC	1983	2	1980
Roofing Products International Kendallville, IN	1983	–	1980
Roofing Systems Inc. St. Clair, MI	1986	–	1980
Schmidt Roofing Co. Miami, OK	1980	1	1980
Seabloom Roofing and Sheet Metal Inc. Toledo, OH	1981	–	1980
Techos Y Derivados Asfalticos Miami, FL	1985	–	1980
Tidwells Urethane Foam Service Inc. Lakeland, FL	1986	2	1980
Tri County Roofing Santa Barbara, CA	1986	–	1980
Tri County Roofing Contractors Robinson, IL	1980	1	1980
Tri-County Roofing Bakersfield, CA	1976	2	1980
TLC Enterprises Jackson, MS	1985	1	1980
US Intec Inc. Port Arthur, TX	1981	–	1980
United Advanced Technologies Inc. Houston, TX	1984	–	1980
United Construction Products Inc. Denver, CO	1983	–	1980

	Joined NRCA	No. of Generations	Founded
United Roofing and Sheet Metal Co., Inc. West Ashville, NC	1981	–	1980
Werle and Sons Inc. Rancho Mirage, CA	1984	2	1980
West Insulation Systems Inc. LaGrange, OH	1986	2	1980
Wheatley Roofing Co., Inc. Louisville, KY	1983	1	1980
Woodall Roofing Inc. Norcross, GA	1981	2	1980
Age Construction and Roofing Co. San Francisco, CA	1984	–	1981
Arizona Roofmaster Inc. Phoenix, AZ	1985	–	1981
Behstev Corp. Fresno, CA	1984	–	1981
Budget Roofing Co. Chicago, IL	1986	1	1981
Commercial Roofing Technology Inc. Fresno, CA	1984	–	1981
Commonwealth Roofing Corp. Louisville, KY	1985	–	1981
Consolidated Industrial Roofing Inc. Roanoke, VA	1984	–	1981
Crowther-Pazely Roofing Co., Inc. Dwight, IL	1982	–	1981
Dimension Roofing Inc. Tucson, AZ	1982	–	1981

	Joined NRCA	No. of Generations	Founded
Empire Roofing Co., Inc. Albuquerque, NM	1986	–	1981
Gerard Tile Co., USA Inc. Orange, CA	1982	–	1981
Jim Giese Commercial Roofing Inc. Dubuque, IA	1981	1	1981
Great Western Roofing Inc. Houston, TX	1984	–	1981
Griffin and Cantrell Co., Inc. Mt. Juliet, TN	1984	–	1981
GLR Roofing Inc. Simi Valley, CA	1986	–	1981
High Plains Roofing Inc. Hays, KS	1986	–	1981
Hoppe Construction Inc. Roseburg, OR	1984	1	1981
Insulation Corp., of America Allentown, PA	1983	–	1981
Johnsons Roof Maintenance Inc. Calhoun, GA	1985	1	1981
Les Jones Roofing Inc. Minneapolis, MN	1981	1	1981
Kline Associated Roofing Contractors Inc. Hagerstown, MD	1982	1	1981
Korellis Roofing Inc. Hammond, IN	1966	2	1981
Lake Shore Roofing and Siding Melrose Park, IL	1986	–	1981
Landmark Single Ply Roofing Systems Atlanta, GA	1984	–	1981

	Joined NRCA	No. of Generations	Founded
The Lathan Co., Inc. Mobile, AL	1986	–	1981
Magnum International Inc. Glenwood, IL	1986	–	1981
Donald Mahon Roofing Inc. Dunellen, NJ	1986	1	1981
R. H. Marcon Inc. State College, PA	1982	1	1981
McKinnis Roofing Blair, NE	1986	1	1981
Metal Air Co. Springfield, IL	1985	–	1981
Metro Roofing and Metal Supply Nashville, TN	1984	–	1981
Ozark Roofing and Guttering Sullivan, MO	1985	2	1981
P and R Roofing and Sheet Metal Inc. Lexington, KY	1983	–	1981
Reinhardt Roofing Inc. San Jose, CA	1986	–	1981
Reliable Roofing and Siding of L.I. Corp. Lindenhurst, NY	1985	–	1981
Rikco Roofing Inc. Baltimore, MD	1981	1	1981
Roofing Southeast Inc. Tampa, FL	1980	–	1981
Roofing Supply Inc. Houston, TX	1984	–	1981
Rooftech Inc. Jeffersonville, IN	1981	1	1981

	Joined NRCA	No. of Generations	Founded
Scanlon Roofing Co. Pomano Beach, FL	1985	–	1981
Sernat Roofing Co., Inc. Ardmore, PA	1983	1	1981
Smith Roofing Co. Tucson, AZ	1984	1	1981
Southwest Products Inc. Sugarland, TX	1984	1	1981
The Tru Fast Corp. Bryan, OH	1982	–	1981
Ucan Fastening Products Inc. Hackensack, NJ	1985	–	1981
Viera's Roofing Contractors Inc. Rio Piedras, PR	1984	–	1981
Weatherproof Services Inc. Hattiesburg, MS	1985	1	1981
Wescoat Corp. Phoenix, AZ	1981	–	1981
West Bend Roofing Inc. West Bend, WI	1984	–	1981
Wolfe Roofing Inc. Calumet Park, IL	1982	–	1981
Wray Roofing Inc. Newton, KS	1985	–	1981
ARM of Texas Inc. Roofing and Sheet Metal Austin, TX	1984	1	1982
AcrySyl International Corp. Reading, PA	1984	–	1982
Allstate Roofing Co., Inc. Memphis, TN	1982	–	1982

	Joined NRCA	No. of Generations	Founded
American Rubber Roofing Inc. Huntertown, IN	1983	–	1982
Architectural Roofing Inc. Phoenix, AZ	1984	–	1982
Associated Roofing Contractors Inc. Owensboro, KY	1983	–	1982
Associated Roofing Services Inc. Birmingham, AL	1984	–	1982
AFĊ Roofing and Insulation Inc. Novi, MI	1984	1	1982
Bishop Wholesale Inc. Martinez, CA	1983	–	1982
Blue Sun Roseville, CA	1985	–	1982
Buck Roofing Co. New Lenox, IL	1984	–	1982
Burmco San Jose, CA	1982	–	1982
Cardinal Roofing Inc. Arlington, TX	1984	–	1982
Colorado Moisture Control Inc. Loveland, CO	1983	–	1982
Crawford Roofing and Maintenance South Holland, IL	1982	1	1982
Dry Bond International Inc. Mundelein, IL	1984	–	1982
Fidelity Roofing Inc. Chicago, IL	1983	2	1982
Finco Roofing Co. Caledonia, MS	1984	1	1982

	Joined NRCA	No. of Generations	Founded
Five Star Roofing San Jose, CA	1986	1	1982
Great Lakes Roofing Inc. Westlake, OH	1986	1	1982
Idelberg and Hayes Denver, CO	1983	–	1982
INRI Inc. Tacoma, WA	1983	1	1982
Jeanco Inc. Fairview, PA	1985	1	1982
Calvin D. Johnson Co., Inc. St. Petersburg, FL	1985	–	1982
Johnston Roofing and Sealcoating Martinez, CA	1985	2	1982
M.J.B. Inc. T/A Built-Up Roofing System Rockville, MD	1983	–	1982
Miami Roofing and Coating Inc. North Miami Beach, FL	1985	–	1982
Mid-Continent Roofing Supply Summit, IL	1982	–	1982
Modi-Systems Inc. Oradell, NJ	1982	–	1982
Moisture Barriers Inc. Altamont, NY	1983	–	1982
Multilite Corp. Coral Gables, FL	1983	–	1982
Myers Roofing Tukwila, WA	1986	1	1982
NTB Fastening Systems Inc. Mosinee, WI	1985	–	1982

	Joined NRCA	No. of Generations	Founded
North American Roofing Co., Inc. Carmel, IN	1982	–	1982
Northeast Roofing Inc. Parma, OH	1986	–	1982
Harry L. Novak AIA San Francisco, CA	1985	–	1982
Palm Royale Roof Corp. St. Petersburg, FL	1984	–	1982
Parma Roofing Inc. Parma Heights, OH	1986	2	1982
Phoenix Roofing Inc. Dallas, TX	1982	–	1982
Posi-Slope Enterprises Inc. Ontario, Canada	1985	–	1982
Pro Co., Inc. Lyman, WY	1986	–	1982
Produits Pour Toitures Quebec, Canada	1985	–	1982
R.L.K. Co., Inc. Tucson, AZ	1955	2	1982
Rawson and Sons Roofing Inc. Omaha, NE	1984	2	1982
Reston Roofing Corp. Herndon, VA	1982	1	1982
Roof Systems Engineering Fresno, CA	1983	–	1982
Roofblok Ltd. Fitchburg, MA	1983	1	1982
Roofers Supply Inc. Cincinnati, OH	1985	–	1982

	Joined NRCA	No. of Generations	Founded
Rubbertite Roofing and Foundation Northwood, IA	1983	–	1982
Schroeder Roofing Inc. Wilmington, NC	1985	–	1982
Scudder Roofing Co. Monterey, CA	1985	1	1982
Seal Dry/USA Inc. Pontiac, MI	1984	–	1982
Show-Me Roof Corp. Centralla, MO	1984	–	1982
Paul Simon Roofers Glen Arm, MD	1984	1	1982
Single Ply Institute of America Pasadena, MD	1983	–	1982
Single-Ply International Inc. Livonia, MI	1983	–	1982
Southern Roofing Inc. Anderson, SC	1982	3	1982
Springer-Peterson Lakeland, FL	1985	2	1982
Standard Roofing of Arizona Inc. Glendale, AZ	1982	–	1982
Styro Systems Inc. Roswell, GA	1982	–	1982
Sunrise Service Corp. Tucson, AZ	1986	–	1982
Surety Contracting Corp. Crawfordville, GA	1984	2	1982
Tecsam Inc. Columbus, OH	1982	–	1982

	Joined NRCA	No. of Generations	Founded
Tennessee Roofing Corp. Knoxville, TN	1983	–	1982
Terstep Roofing Inc. Noblesville, IN	1982	–	1982
Mike Tighe Roofing Inc. Fond du Lac, WI	1985	2	1982
Unique Inc. Rockville, MD	1985	–	1982
Vascocu's Roofing and Sheet Metal Inc. Baton Rouge, LA	1984	1	1982
Weathercraft Co. of Colo. Springs Colorado Springs, CO	1983	–	1982
WestPoint-Pepperell West Point, GA	1982	–	1982
A and R Roofing Inc. Southfields, NY	1985	–	1983
Aceves Construction and Maintenance Co. Norfolk, VA	1984	–	1983
Albuquerque Equipment and Roofing Supply Albuquerque, NM	1983	2	1983
Alderson Roofing and Sheet Metal Co., Inc. Memphis, TN	1986	1	1983
Apple Roofing Corp. Syracuse, NY	1986	–	1983
Armko Industries Inc. Dallas, TX	1984	–	1983
Bell Roofing and Maintenance Inc. Trenton, NJ	1986	–	1983

	Joined NRCA	No. of Generations	Founded
Brandle Roofing and Sheet Metal Co. Midland, MI	1984	1	1983
Jerry Colbert Roof Service Springfield, VA	1983	–	1983
Conway Roofing Co., Inc. North Billerica, MA	1985	–	1983
CDA International Inc. State College, PA	1986	–	1983
Diamond Roofing Co. Dallas, TX	1985	–	1983
Diversified Roofing and Insulation Co. Bedford, OH	1983	–	1983
Dostoomian Roofing Systems Inc. North Abington, MA	1985	–	1983
Doxsee Roofing Inc. Bennington, VT	1984	–	1983
Eagle Roofing Inc. South Point, OH	1985	–	1983
Fisher/Hart Enterprises LaGrange, IL	1985	–	1983
Paul Frank Roofing and Waterproofing Corp. Philadelphia, PA	1985	–	1983
Harrington Roofing Co. Kansas City, KS	1984	1	1983
Higgins and Higgins Inc. Scarsdale, NY	1984	2	1983
High Tech Roofing Co., Inc. Stoneham, MA	1984	–	1983

	Joined NRCA	No. of Generations	Founded
A. D. Holst Roofing and Sheet Metal Inc. Cedar Rapids, IA	1985	–	1983
Howard and Ruscilli Inc. Dayton, OH	1984	–	1983
Industrial Contractors Omaha, NE	1985	–	1983
Industrial Energy Systems Strongsville, OH	1984	–	1983
Industrial Roofing and Sheeting Co. Johnstown, PA	1985	–	1983
Island Roofing Corp. Ronkonkoma, NY	1986	–	1983
Jamco Inc. Decatur, GA	1985	1	1983
Jorve Roofing Co. Seattle, WA	1986	1	1983
Kantron Roofing Corp. Astoria, NY	1984	–	1983
Lower Penisula Roofing and Sheet Metal Inc. Detroit, MI	1985	–	1983
M and W Construction Systems Inc. Elkridge, MD	1985	–	1983
Marion and Green Roofing and Sheet Metal Co. LaVergne, TN	1984	–	1983
Midwest Roofing and Sheet Metal Inc. Tea, SD	1984	–	1983

	Joined NRCA	No. of Generations	Founded
Midwestern Roofing Distributors Tremont, WI	1986	–	1983
Modern Roofing Inc. Lynn, MA	1985	–	1983
Nevco Inc. Houston, TX	1985	–	1983
Oklahoma Roofing and Sheet Metal Oklahoma City, OK	1985	–	1983
Pardo Roofing Glen Head, NY	1985	1	1983
Power Marketing Group Inc. Denver, CO	1983	2	1983
Quality Roofing Inc. Bradford, CT	1986	–	1983
The Roof Co. Cheyenne, WY	1985	–	1983
Roof Inspection Service Sewickley, PA	1985	–	1983
Roof Systems of The Piedmont Easley, SC	1984	3	1983
Roofers Mart of Georgia Smyrna, GA	1983	–	1983
Roofing Systems Inc. Kent, WA	1986	–	1983
Stricker Roofing Inc. Escondido, CA	1985	1	1983
Strickland Builders Golden, MS	1985	–	1983
Talon Roofing and Waterproofing Inc. Round Rock, TX	1985	2	1983
Technological Roof Systems Inc. McLean, VA	1984	–	1983
Tomlin Corp. Addison, IL	1985	–	1983
Weathercraft Co. of Grand Island Grand Island, NE	1984	–	1983
Weatherguard Inc. Dallas, TX	1983	1	1983
White Roofing and Waterproofing Inc. San Antonio, TX	1983	–	1983
Wilson Cover Co., Inc. Wilson, NC	1985	2	1983
AIP International Corp. Northvale, NJ	1986	–	1984
A-Top Roofing Inc. Brea, CA	1986	2	1984
R. Adams Roofing Inc. Indianapolis, IN	1986	1	1984
Altec Roofing Inc. Jupiter, FL	1984	–	1984
Barge Terminal and Trucking Inc. Chicago, IL	1985	–	1984
Bonner Roofing and Metal Inc. Lufkin, TX	1985	–	1984
Burlington Industries Inc. Greensboro, NC	1984	–	1984
Caribbean Systems St. Thomas, VI	1986	–	1984
Carpenter Insulation Co. Richmond, VA	1985	–	1984

	Joined NRCA	No. of Generations	Founded
Cooperative Roofing Inc. Casselberry, FL	1985	–	1984
County Line Contractors Inc. Ft. Lauderdale, FL	1986	–	1984
Flint Regional Roofing Inc. Burton, MI	1986	2	1984
Foothills Roofing and Sheet Metal Inc. Easley, SC	1984	–	1984
The Gehringer Corp. Allentown, PA	1986	–	1984
Greenbriar Building Materials Inc. Pompano Beach, FL	1985	–	1984
GME Consultants Inc. Bridgeview, IL	1986	1	1984
Horace L. Heath Inc. Lavergne, TN	1984	–	1984
Insul-Mark Midwest Inc. LaPorte, IN	1986	1	1984
Kansas City Roofing Center Inc. Kansas City, MO	1984	–	1984
Knost Roofing and Construction Co., Inc. Shreveport, LA	1985	–	1984
L and H Enterprises Inc. Annapolis, MD	1986	–	1984
Laufenberg and Sons Roofing and Sheet Metal Inc. Waukesha, WI	1985	1	1984
Lindholm Roofing Chicago, IL	1985	2	1984
Marion/Service Roofing and Air Conditioning Co. Ocala, FL	1984	–	1984
National Roofing of Kansas Inc. Hays, KS	1986	1	1984
Nu-Tec Roofing Contractors Inc. Indianapolis, IN	1984	–	1984
Petry Brothers Inc. Urbana, IL	1984	–	1984
Douglas S. Plotke Jr. Inc. Islip, NY	1984	–	1984
Rocky Mountain States Roofers Mart Denver, CO	1985	–	1984
Roofers Mart of Southern California Inc. Walnut, CA	1985	–	1984
Roofers Mart of America Inc. St. Louis, MO	1984	–	1984
Roofers Mart of Central Pennsylvania Lancaster, PA	1984	–	1984
Roofers Mart of Central Texas Inc. San Antonio, TX	1984	–	1984
Roofers Mart of Wisconsin Inc. Milwaukee, WI	1984	–	1984
Santa Ana Roofing Co. Santa Ana, CA	1986	–	1984
Simon Roofing and Sheet Metal Corp. Youngstown, OH	1986	3	1984

	Joined NRCA	No. of Generations	Founded
Single Ply Systems Inc. Alsip, IL	1986	–	1984
Soderstrom Architects Portland, OR	1986	–	1984
Southeastern Roofing and Siding Inc. Virginia Beach, VA	1986	–	1984
Sprinkel and Associates Los Altos, CA	1986	–	1984
U-Flow Inc. Buffalo, NY	1984	1	1984
Ultra Roof Co. Kent, WA	1986	–	1984
Weather-Tek Building Prod. Hales Corners, WI	1985	–	1984
Weathermaster Roofing Co., Inc. Binghamton, NY	1984	–	1984
Western Pacific Ventures Anchorage, AK	1986	1	1984
Accredited Contractors Clinton, MD	1986	–	1985
Allstate Roofing Inc. Lenexa, KS	1986	–	1985
The Beck Co. Cleveland, OH	1986	2	1985
Bill's Roofing Service Inc. Worthington, MN	1985	–	1985
Capital Roofing Systems Inc. Crestwood, IL	1986	–	1985
Cascade Roofing and Sheet Metal Inc. Seattle, WA	1985	–	1985

	Joined NRCA	No. of Generations	Founded
F and R Roof Maintenance Inc. Johnstown, NY	1985	–	1985
Grabman Associates Grand Haven, MI	1986	–	1985
Gravel Vac Systems Inc. Stoneham, MA	1985	1	1985
Hite Associates Cumberland, MD	1986	–	1985
Humboldt Roofing Co. Sparks, NV	1986	2	1985
Impex Inc. Hurst, TX	1984	–	1985
Innovative Roofing Systems Inc. Salt Lake City, UT	1986	2	1985
International Coating Systems Las Vegas, NV	1986	1	1985
K and B Roofing Inc. Laurel, MD	1986	1	1985
Lage Roofing Jackson Hole, WY	1985	–	1985
Robert E. Linck Inc. Philadelphia, PA	1985	–	1985
M and M Roofing Co., Inc. Duluth, GA	1986	–	1985
R. F. Prucnal Roofing Co. Kaneohe, HI	1985	–	1985
Renaissance Roofing Services Inc. Greenfield, WI	1986	–	1985
Roofing Concepts Unlimited Ft. Lauderdale, FL	1985	–	1985
Technological Roof Services Inc. Pleasant Hill, CA	1985	–	1985

	Joined NRCA	No. of Generations	Founded
Texas Roofmasters Dallas Inc. Dallas, TX	1986	–	1985
Tuff-Con Phoenix Phoenix, AZ	1986	–	1985
A-Tech Fastener Corp. West Springfield, MA	1986	–	1986
Hardcore Roofing Inc. Merrionette Park, IL	1986	–	1986
Imperial Roofing Co., Ltd. New Ipswich, NH	1986	1	1986
Industrial Roofing Services Inc. Racine, WI	1986	–	1986

	Joined NRCA	No. of Generations	Founded
Meta Inc. Texarkana, TX	1986	–	1986
Metropolitan Restoration and Waterproofing Dorchester, MA	1986	–	1986
Newtson Roofing Co., Inc Ottawa, IL	1986	–	1986
Saturn Associates Schaumburg, IL	1986	–	1986
Star Manufacturing Co. Oklahoma City, OK	1986	–	1986

1890: McKinley Tariff passed. Fixes duty of 2.2 cents per pound on imported tin-plate, dramatically increasing production and popularity as roofing material in the U.S.

1891: Philadelphia City Hall roofed with aluminum.

1893: Terneplate selected to cover all buildings in Chicago World's Fair.

1902: Under guidance of American Federation of Labor, United Slate, Tile and Composition Roofers Damp and Waterproofing Workers Association is formed.

1906: Barrett Specifications published, which company calls "bible of good roofing." Minimal guidelines for slag and gravel roofs included.

1911: Prepared Roofing Manufacturers Association formed, forerunner of ARMA.

1911: *American Roofer* magazine founded by Eugene M. Pope, secretary of National Association of Master Composition Roofers.

1916: Based on Underwriters Laboratories testing, National Board of Fire Underwriters assigns roof coverings to classes A through H.

1916: National Board of Fire Underwriters calls for ban on wood shingles. Asphalt shingle market is born.

1916: Barrett manufactures complete roofing system—Barrett System Roof—and backs it with first bond. Company starts qualifying "Barrett Approved Roofers" to apply materials.

1916: Associated Roofers of America formed.

1921: ARA joins the National Association to form United Roofing Contractors Association of North America (URCA).

1926: Prepared Roofing Manufacturers Association changes name to the Asphalt Shingle and Roofing Association.

1929: Stock market crashes.

1933: Construction expenditures plunge to $3 billion from high of $12.5 billion in 1926.

1934: Eugene Pope dies. Harriet B. Wagner named secretary of Association.

1935: Northeastern Roofing, Siding and Insulation Contractors Association formed. James McCawley, a New York roofing contractor, named executive director and starts *Modern Roofer* magazine, which combines with *American Roofer* to become the *American Roofer and Modern Roofing.*